D0915722

WITHDRAWN

The Poverty of the Linnaean Hierarchy

The question of whether biologists should continue to use the Linnaean hierarchy is a hotly debated issue. Invented before the introduction of evolutionary theory, Linnaeus's system of classifying organisms is based on outdated theoretical assumptions and is thought to be unable to provide accurate biological classifications.

Marc Ereshefsky argues that biologists should abandon the Linnaean system and adopt an alternative that is more in line with evolutionary theory. He traces the evolution of the Linnaean hierarchy from its introduction to the present. Ereshefsky illustrates how the continued use of this system hampers our ability to classify the organic world, and then he goes on to make specific recommendations for a post-Linnaean method of classification.

Accessible to a wide range of readers by providing introductory chapters to the philosophy of classification and biological taxonomy, the book will interest both scholars and students of biology and the philosophy of science.

Marc Ereshefsky is Associate Professor of Philosophy at the University of Calgary.

The Poverty of the Linnaean Hierarchy

A Philosophical Study of Biological Taxonomy

MARC ERESHEFSKY

University of Calgary

PUBLISHED BY THE PRESS SYNDICATE OF THE UNIVERSITY OF CAMBRIDGE
The Pitt Building, Trumpington Street, Cambridge, United Kingdom

CAMBRIDGE UNIVERSITY PRESS
The Edinburgh Building, Cambridge CB2 2RU, UK
40 West 20th Street, New York, NY 10011-4211, USA
10 Stamford Road, Oakleigh, VIC 3166, Australia
Ruiz de Alarcón 13, 28014 Madrid, Spain
Dock House, The Waterfront, Cape Town 8001, South Africa

http://www.cambridge.org

First published 2001

Printed in the United States of America

Typeface Times Roman 10.25/13 pt.   *System* QuarkXPress [BTS]

A catalog record for this book is available from the British Library.

Library of the Congress Cataloging in Publication data

Ereshefsky, Marc.
The poverty of the Linnaean hierarchy : a philosophical study of biological taxonomy /
Marc Ereshefsky.
p. cm. – (Cambridge studies in philosophy and biology)
Includes bibliographical references and index.
ISBN 0-521-78170-1 (hardcover)
1. Biology – Classification – Philosophy.   I. Title.   II. Series.
QH83 .E73 2000
578′.01′2 – dc21   00-028945

ISBN 0 521 78170 1   hardback

Contents

Contents

Preface

It is not surprising that biologists and philosophers wonder about the nature of species. At first glance we feel assured that we know what we are talking about when it comes to species, but when we take a closer look, matters get more complicated and less obvious. I started my research on species first as a graduate student with Elliott Sober and then as a post-doctorate fellow with David Hull. Both taught me how metaphysics applied to biology can be a satisfying and rewarding form of philosophy.

Shortly after my graduate studies, I started thinking more generally about the nature of species. Instead of worrying about their proper biological description or their ontological status, I started to wonder about their role in evolutionary theory. Experts told me that species are units of evolution. I looked at that notion and found its meaning ambiguous and often vague. Perhaps a better understanding of the distinction between species and other types of taxa (genera, families, and so forth) would help. The deeper I dug, the more problems appeared – the distinctions among those types of taxa were far from clear. Soon it became apparent that the entire Linnaean hierarchy of categorical ranks had dubious theoretical underpinnings. What about the procedures we use for naming taxa, since they stem from Linnaeus's system of classification as well? Again, problems began to surface. The more closely I looked at our current system of classification, the more skeptical I became of its ability to represent the organic world. This book summarizes my investigation of the Linnaean system. It shows why I believe the Linnaean system's days are numbered.

Many people helped me write this book. My views on biology and classification have been shaped by numerous conversations and correspondences. For those, I thank David Baumslag, Kevin de Queiroz,

John Dupré, Mohamed Elsamahi, Berent Enc, Michael Ghiselin, Nick Jardine, Jim Lennox, David Magnus, Charlie Martin, Ernst Mayr, David Mellow, Mark Migotti, Brent Mishler, Jay Odenbaugh, Janet Sisson, Ken Waters, Bradley Wilson, and Brad Wray. I would especially like to thank David Hull, Allan Larson, Michael Ruse, Tony Russell, and Elliott Sober. Each read the entire manuscript or most of it, and each provided me with pages of insightful, challenging, and useful comments. I cannot thank them enough for their efforts, as well as their personal support.

I received help from many other folks. Bruce Collins and Barrett Wolski were research assistants early on in this project. Gwen Seznec and Robyn Wainner at Cambridge University Press suffered through my naïve questions concerning publishing. My home department – the Department of Philosophy at the University of Calgary – has been a warm and encouraging environment for doing philosophy. The University of Calgary provided me with time to write, first in the form of a sabbatical leave and later as a fellow at the Calgary Humanities Institute. Further funding for research on this project was provided by the National Science Foundation, the Isaak Walton Killam Memorial Fund, and the Center for Philosophy of Science at the University of Pittsburgh.

Some of the material used in this book has been published elsewhere. I thank Robert Sokal and Peter Sneath for letting me reprint their diagram "A Flow Chart of Numerical Taxonomy" from *Principles of Numerical Taxonomy*. The University of Illinois Press granted permission to use Figures 6 and 18 of Willi Hennig's *Phylogenetic Systematics*. Kluwer Academic Publishers allowed me to use portions of my "The Evolution of the Linnaean Hierarchy," *Biology and Philosophy* 12/4 (1997). The frontispiece of Linnaeus's *Hortus Cliffortianus* reproduced in the introduction to this text appears courtesy of the Hunt Institute for Botanical Documentation, Carnegie-Mellon University, Pittsburgh.

Finally, I would like to thank Ayala and Josh for keeping me happy and sane during the five years that I worked on this book. As always, Ayala's support and encouragement kept me going. Last, I would like to thank my parents, Philip and Rachel. This book is for them.

Introduction

THE LINNAEAN SYSTEM RECONSIDERED

Three hundred years ago biological taxonomy was a chaotic discipline marked by miscommunication and misunderstanding. Biologists disagreed on the categories of classification, how to assign taxa to those categories, and even how to name taxa. Fortunately for biology, Linnaeus saw it as his divinely inspired mission to bring order to taxonomy. The system he introduced offered clear and simple rules for constructing classifications. It also contained rules of nomenclature that greatly enhanced the ability of biologists to communicate. Linnaeus's system of classification was widely accepted by the end of the eighteenth century. That acceptance brought order to a previously disorganized discipline. Furthermore, it laid the foundation for "the unprecedented flowering of taxonomic research" of the late eighteenth and early nineteenth centuries (Mayr 1982, 173).

Linnaeus himself seemed assured of his place in the history of biology. Consider the frontispiece of his *Hortus Cliffortianus* (1737) (Figure I.1). Linnaeus's youthful face is seen on the body of Apollo. In one hand he holds a light, in the other he pushes back the clouds of ignorance from crowned Mother Nature. With his foot Linnaeus tramples the dragon of falsehood. In the foreground, plants are brought for identification and two cherubs admire Linnaeus's centigrade thermometer. An exuberant illustration, and an immodest one – it was commissioned and approved by Linnaeus. The metaphors of the illustration are not completely unfounded; Linnaeus's work did usher in a golden era of biological classification.

In the last two hundred years, the theoretical landscape of biology has changed drastically. The foundation of Linnaeus's system was his

1

Figure I.1   The frontispiece of Linnaeus's *Hortus Cliffortianus* (1737). Courtesy of Hunt Institute for Botanical Documentation, Carnegie Mellon University, Pittsburgh, PA.

biological theory – in particular, his assumptions of creationism and essentialism. These assumptions have gone by the wayside and have been replaced by evolutionary theory. Still, the vast majority of biologists use the Linnaean hierarchy and its system of nomenclature. Unfortunately, the system's outdated theoretical assumptions undermine its ability to provide accurate classifications. Furthermore, its rules of nomenclature, once prized for their ability to bring order to biological classification, are no longer practical. These problems are far from minor, for the Linnaean system is the backbone of biological classification and much of biology. The Linnaean system prescribes how to name and represent taxa and, in doing so, provides the template for displaying life's diversity. Moreover, the terms and concepts of the Linnaean system play a central role in biological theorizing. They frame all theoretical questions concerning groups of organisms above the level of the local population.

To get a better idea of the problems facing the Linnaean system, let us take a closer look at its theoretical assumptions. Among them is Linnaeus's conception of biological taxa. Linnaeus thought that species and other taxa are the result of divine intervention. Once a taxon is created, each of its members must have the essential properties of that taxon. The evolution of a species was foreclosed by God's original creation. Needless to say, Darwinism gives us a different picture of the organic world. Taxa are the products of natural rather than divine processes. Species are evolving lineages, not static classes of organisms. This conceptual shift in biological theory is well discussed in the literature and comes under many banners. Some authors talk of the "death of essentialism"; others refer to the "species are individuals" thesis. In broader perspective, this conceptual shift falls under "the historical turn" in biology, or what Ernst Mayr calls "population thinking."

Essentialism concerning taxa has fallen out of favor among evolutionary biologists, so this tenet of Linnaeus's original system has been dropped. Nevertheless, many of Linnaeus's original principles remain in place, and those assumptions, I will argue, are equally problematic. Consider Linnaeus's conception of the species category. The species *category* is the group of all species taxa, whereas species *taxa*, such as *Homo sapiens* and *Drosphila melangaster*, are groups of organisms. For Linnaeus, not only do species taxa have essences, but so does the species category. In the Linnaean system, all species taxa are comparable and distinguishable from all other types of taxa. The assumption that there is an essence to the species category is still widely held. Many biologists believe that species are groups of organisms that can successfully interbreed and produce fertile offspring. Being a group of organisms with those properties defines membership in the species category. So while biologists have rejected the Linnaean assumption that species taxa have essences, they have, for the most part, retained the assumption that there is an essence to the species category.

That latter assumption, however, should be questioned as well. The idea that species are groups of organisms that successfully interbreed and produce fertile offspring is just one of many prominent definitions of the species category. (Biologists often refer to such definitions as "species concepts.") Another definition asserts that a species is a group of organisms bound by their unique phylogeny, and still another defines a species as a group of organisms that share a unique ecological niche. What are we to make of this variety of species concepts in the literature? According to some authors, there is a single correct description of the species category. The existence of more than one species concept, they suggest, merely reflects a lack of consensus among biologists on the nature of species. Against this view, I will argue that there is no single correct definition of the species category. The species category lacks an essence and is in fact multifarious.

This second view is species pluralism and runs counter to the Linnaean assumption that there is an essence to the species category. If one accepts species pluralism, then not only must essentialism at the level of species be abandoned, but so must essentialism at the level of the species category. Similar arguments have been used against the other Linnaean categories. Many biologists note that the higher Linnaean categories – genus, family, order, and so forth – are heterogeneous collections of taxa. Families, for example, vary in their ages as well as their degrees of inclusiveness. Calling a taxon a "family" indicates only that within a particular classification that taxon is more inclusive than genera and less inclusive than classes. This meaning of "family" has no ontological significance. Similar observations apply to the rest of the Linnaean higher categories. If the higher Linnaean categories lack significant defining features, then another major tenet of the Linnaean system is obsolete.

Biologists and philosophers have carefully examined the fate of essentialism concerning species taxa. However, very little attention has been paid to essentialism at the level of the Linnaean categories. We should look more carefully at the assumption that the Linnaean categories have essences, not only because that assumption may be unfounded, but also because it underwrites many practices prescribed by the Linnaean system.

Consider the project of finding *the* definition of the species category, or the practice of finding *the* correct rank of a taxon. Suppose, as suggested, that the Linnaean categories, including the species category, are heterogeneous collections of taxa. If the taxa of a particular Linnaean category are not comparable on some significant parameter, then we should wonder about that category's existence. If the Linnaean categories do not exist, then the Linnaean ranks are ontologically empty designations. Nevertheless, we still assign taxa such ranks and, for the most part, think that those designations have significance. All species taxa, it is often assumed, occupy a unique and common role in the economy of nature. But we may have labored under a false assumption. Being wrong is bad enough, but when inappropriate theoretical commitments lead to practices that waste valuable research time, that mistake is compounded. If species taxa do indeed form a class of heterogeneous entities, then time and energy is wasted when we argue over *the* correct definition of the species category. Similarly, the assumption that the Linnaean categories represent real levels of genealogical inclusiveness in nature encourages biologists to argue over the rank of taxa; yet such disagreements may lack an objective basis for resolution. Accordingly, Willi Hennig complains that the continued use of the Linnaean categories is the source of many "unfruitful debates" in biological taxonomy (1969, xviii).

The ontological problems with the Linnaean hierarchy lead to further practical problems. The current Linnaean system contains a number of rules of nomenclature, many of which have their source in Linnaeus's original system. A centerpiece of those rules is the requirement that the Linnaean ranks of taxa be incorporated in taxon names. Species are given binomials, whereas more inclusive taxa are assigned uninomials. In addition, the names of many higher taxa have rank-specific endings. Unfortunately the ontological problems of the Linnaean categories undermine these rules of nomenclature. Suppose, as suggested, that the existence of the Linnaean categories should be doubted. We might then wonder if it is appropriate to indicate a taxon's rank in its name when such ranks correspond to nothing in nature. If there is no species category or genus category, then no taxon should be designated as a species or a genus. This is not merely a case in which certain scientific concepts are idealized representations that nonetheless approximate the natural world. The Linnaean categories may have no basis in nature.[1]

There are other problems with the Linnaean rules of nomenclature regardless of whether one is skeptical of the Linnaean categories. The simple requirement that a taxon's name indicate its rank causes needless instability in biological taxonomy. Classifications of the organic world are constantly revised in light of new evidence, and often such revisions require that the ranks of taxa be altered. The Linnaean system makes such revisions doubly hard. Biologists must change not only the taxonomic positions of taxa but also their names. The need to revise classifications is an epistemological problem that cannot be eliminated from biological taxonomy. Nevertheless, we would like classifications to remain as stable as possible. The instability of a taxon's name can be avoided if we drop the Linnaean requirement that a taxon's name contain information about its taxonomic position. The Linnaean rules of nomenclature lead to further practical problems. For instance, when biologists disagree on the rank of a taxon they are required by that system to give two different names (each indicating a different rank) to what they agree is the very same taxon. The list of problems continues.

Stepping back, the following can be said of the continued use of the Linnaean system. Taxonomists and philosophers of biology are familiar with the death of essentialism concerning species taxa, but that change in thought concerns only a small portion of the Linnaean system. Many problematic aspects of that system remain in place and continue to guide the vast majority of taxonomists in constructing classifications. Given the problems facing the Linnaean system and that system's importance in biology, a philosophical investigation of the Linnaean system is sorely needed. The aim of this book is to provide that analysis. We will begin with such broad issues as the nature of scientific classification and eventually work our way to specific recommendations for a post-Linnaean system. In an effort to give a clearer idea of what is to come, the remainder of this introduction provides a survey of the book's chapters.

An Overview of the Book

Starting at a rather global level, *The Poverty of the Linnaean Hierarchy* explores various philosophical problems in biological taxonomy. Many of these issues appear in both biology and philosophy, though under different guises. Philosophers would like to know how best to represent the world's entities, what the relationship is between theory and classification, and other such questions. Biologists are interested in the same questions, but in the more restricted domain of biology. This book examines how the quest for scientific classifications has occurred in biology by looking at the Linnaean system of classification. But before getting to the details of biological taxonomy itself, we need an introduction to some of the issues in the philosophy of classification. Chapter 1 outlines various philosophical approaches to classification, from essentialism to Wittgenstein's notion of family resemblance to more recent suggestions of homeostatic cluster kinds. Because much of this book is devoted to the historical turn in biological taxonomy, a large part of Chapter 1 discusses the notion of historical classification and the identity and individuation of historical entities. It is one thing to say that a particular taxon is a historical entity; it is another to give precise conditions for its identity. Other debates in the philosophy of classification are surveyed, such as the debate between monists and pluralists and the debate between realists and anti-realists. Chapter 1 provides a menu of options one can choose from in developing a philosophy of classification.

Chapter 2 turns to biology proper and is a primer of biological taxonomy. All active scientific disciplines contain theoretical disagreements. The same is true of biological taxonomy, though the disagreements there are more properly described as disagreements over the appropriate methods for developing classifications. Contemporary biology contains no fewer than four general schools of taxonomy: evolutionary taxonomy, pheneticism, process cladism, and pattern cladism.[2] When one turns to the literature on the nature of species, the number of options increases. The first half of Chapter 2 introduces the major schools of biological taxonomy, while the second half introduces six prominent species concepts. The debate among biologists over the proper school of taxonomy has been heated and at times rancorous. I will try to stay out of the fray in this chapter (though I will take sides in later chapters). Chapter 2 merely provides the biological background for the philosophical problems discussed in later chapters. For those well acquainted with contemporary biological taxonomy, the material in Chapter 2 will be familiar. For those lacking a strong background in the field, Chapter 2 provides important information for understanding the topics discussed in later chapters.

An issue that has come to the fore in recent years is the ontological status of biological taxa, especially species. Many have argued that a Darwinian view of biological taxa requires that we treat taxa as historical entities rather than as classes of similar organisms. Traditionally, philosophers as well as biologists have treated taxa as qualitative kinds, where membership turns on the organisms of a taxon being similar to one another in one or more respects. Chapter 3 demonstrates why qualitative approaches to classification, from essentialism to Richard Boyd's homeostatic cluster kinds, fail to provide adequate accounts of biological taxa. Instead, a historical approach, one that views taxa as wholes whose parts (organisms) are causally connected, should be adopted. One component of the historical approach is the "species are individuals" thesis. Much ink has been spilled in the debate over whether species are individuals, yet key elements of that debate remain unresolved. For one, the term "individual" is ambiguous and that ambiguity is a source of disagreement and confusion. Chapter 3 offers an analysis of individuality that disambiguates that notion. It also makes clear what is and is not significant about the claim that species are individuals. Some authors have argued that the "species are individuals" thesis has broad and profound implications for evolutionary theory. The message I will convey is that the historical turn in biological classification does have important ramifications for constructing biological classifications. However, the alleged broader implications of the individuality thesis – for example, that it affects the nature of evolutionary theory as a scientific theory – are not forthcoming.

Having dealt with the question of essentialism at the level of taxa, we turn to the question of essentialism at the level of the species category. Should we assume that there is a single correct definition of the species category, as essentialism dictates, or should we allow that there might be a number of different types of species taxa? Species monism is the traditional view, stemming from essentialism. Species pluralism parts with that view and maintains that there is no essence to the species category. The job of Chapter 4 is to provide a comprehensive argument for the acceptance of species pluralism. Reduced to its barest bones, the argument of Chapter 4 suggests that species pluralism reflects the fecundity of nature, not our lack of understanding of the organic world. Biological forces, it is argued, cause the existence of different types of base taxa. The claim here is ontological, and it is quite different from the epistemological argument that we should prefer pluralism because we lack sufficient evidence to choose one species concept over another. The advocacy of species pluralism is not new to this book, but the arguments presented here differ from those previously found in the literature.

Monists are not happy with the advocation of species pluralism. One objection raised by monists stands out and deserves special attention. Pluralists, they argue, fail to provide adequate criteria for determining which species concepts are worthy of acceptance. Without such criteria, species pluralism allows the acceptance of any suggested classification. Some monists conclude that species pluralism boils down to a position of "anything goes," a position that poses an important methodological challenge to pluralists. Chapter 5 attempts to meet that challenge. We could reflect philosophically on which criteria should be used for selecting species concepts; but we might end up with criteria that are irrelevant to the aims of working taxonomists. A better way is to establish the aims of biological taxonomy, according to biological taxonomists, and then determine which criteria pick species concepts that best satisfy those aims. Larry Laudan calls this approach to scientific methodology "normative naturalism." We derive rules for picking theories, or species concepts, according to the aims of the discipline at hand. The word "naturalism" is used to contrast this sort of philosophy of science from one that determines criteria for selecting theories on just the basis of conceptual, or armchair, analysis. With such criteria in hand, pluralists can say which taxonomic approaches should be accepted and which should be rejected; they can show that pluralism is not vulnerable to the "anything goes" objection, but a position with careful checks in place.

The discussion of species pluralism leads us back to more general questions about the continued use of the Linnaean hierarchy in biology. If the organic world consists of different types of species taxa, and there is no parameter common to species taxa that distinguishes them from other types of taxa, then the species category has lost much of its significance. Similar concerns affect the status of the other Linnaean categories. If the Linnaean categories lack an ontological foundation, then an important tenet of the Linnaean system, a tenet carried to this day, has been lost.

For this reason, and those mentioned earlier, a full scale examination of the Linnaean system should be conducted. The rest of this book is devoted to that task. The first order of business is an introduction to Linnaeus's original system and his motivations for that system. We then turn to the evolution of the Linnaean system, starting with the Darwinian revolution, working through the evolutionary synthesis, and concluding with recent cladistic amendments. Though much has been written on Linnaeus's original system, little has been written on the subsequent evolution of that system. Chapter 6 sketches the transformation of the Linnaean system since its inception.

In the last fifty years, two versions of the Linnaean system have come to the fore. One is the traditional system offered during the evolutionary synthesis by Ernst Mayr and Gaylord Simpson. The other is Edward Wiley's annotated Linnaean system. Wiley's system is a cladistic version of the Linnaean system designed to overcome the problems of the traditional system. Wiley's system does avoid some of the problems facing the traditional system. However, many problems of the traditional system are carried over to Wiley's system. Moreover, Wiley's annotated Linnaean system brings its own problems. Chapter 6 introduces the problems facing traditional and cladistic versions of the Linnaean system. Chapter 6 also suggests why the Linnaean system remains entrenched in biological taxonomy despite its lack of a theoretical foundation.

The problems facing the Linnaean system are significant enough for us to consider the possibility of adopting a non-Linnaean system of classification. But before doing that, we need an alternative system. Saying that the Linnaean system is flawed is one thing; providing a compelling argument for its replacement is quite another. An important step in establishing that the Linnaean system should be replaced is developing an alternative system. A number of post-Linnaean systems have been suggested in the last thirty years. Some proposals address new ways of displaying hierarchical relations, others discuss alternative methods for devising taxon names, still others provide non-Linnaean means for defining taxon names. Unfortunately, alternatives to the Linnaean system tend to be piecemeal, and they contradict one another. Chapter 7 attempts to bring order to this literature. First it examines various proposals for post-Linnaean taxonomy, then it selects the best of those proposals and weaves them together into a coherent alternative system. Again, the motivation here is to provide a viable alternative system to the Linnaean system, because without such an alternative, recommendations for dropping the Linnaean system will go nowhere.

With a post-Linnaean system in hand, we can compare the Linnaean system to a comprehensive alternative. If such a comparison is to be fair, we must contrast the post-Linnaean system with the best developed version of the Linnaean system. That version is Wiley's (1979,

1981) annotated Linnaean hierarchy. The first half of Chapter 8 compares the annotated Linnaean system with the post-Linnaean system developed in the previous chapter. The result is a scoreboard highlighting the desirability of each system. The second half of Chapter 8 steps back from a detailed examination of the annotated and post-Linnaean systems and considers a more general issue. Suppose the post-Linnaean system better coheres with evolutionary theory and, if adopted, would make the job of taxonomists easier. Suppose, in other words that the post-Linnaean system is preferable for both theoretical and pragmatic reasons. At first glance such reasons seem sufficient for adopting a new system of nomenclature, but in practice they are not. Given the pervasiveness of the Linnaean system both in and outside of biology, we need to show that the switch to an alternative system is practically feasible. More precisely, we need to show that the entrenchment of the Linnaean system does not foreclose the possibility of adopting an alternative system.

The Linnaean system has served biology well. Few would deny that. But biological theory has changed drastically in the last two hundred years. Those changes have rendered the Linnaean system both theoretically outdated and pragmatically flawed. One aim of this book is to show that biology is no longer well served by the Linnaean system. Another aim is to show that the replacement of the Linnaean system should be seriously considered.

I

The Historical Turn

1

The Philosophy of Classification

"The philosophy of classification" is not a phrase one frequently comes across in either philosophy or biology. Yet the philosophy of classification addresses foundational issues in both disciplines. Consider its two main questions: How should we classify the world's entities? What is the relationship between classification and the world itself? Philosophers interested in ontology, epistemology, and representation wrestle with both questions. Biologists, particularly biological taxonomists, confront the same questions when considering how to construct classifications of the organic world.

This chapter provides an introduction to the philosophy of classification by starting with the first of the above questions: How should we classify the world's entities? The focus of this book is classification in science, so we need to ask how scientific classifications should be constructed. The history and philosophy of science offer numerous answers. To tackle the wealth of suggested approaches, three general philosophical schools will be presented: essentialism, cluster analysis, and historical classification. Essentialism sorts entities according to their essential natures. Cluster analysis divides entities into groups whose members share a cluster of similar traits, though none of those traits are essential. The historical approach classifies entities according to their causal relations rather than their intrinsic qualitative features. The job of the first three sections of this chapter is to distinguish these major approaches to classification. Section 1.1 introduces essentialism, Section 1.2. outlines various cluster approaches, and Section 1.3 is an extended discussion of the historical approach.

The introductions to these approaches will be unequal. Much more time will be spent on the historical approach than on the qualitative approaches of essentialism and cluster analysis. One reason is that

15

much of this book is devoted to a detailed analysis of the application of the historical approach in biological taxonomy. Thus it requires a more thorough introduction. Another reason is the lack of attention philosophers have paid to the historical approach. Philosophers have focused primarily on qualitative approaches to classification and given little attention to the historical approach. Section 1.3 takes the time to address some foundational issues in historical classification.

The possibility of taxonomic pluralism is another issue explored in this volume. Monists prefer a single classification of a discipline's entities. Pluralists allow a number of equally acceptable classifications of those entities. Taxonomic pluralism itself has various forms (as does taxonomic monism). And the question of taxonomic pluralism raises such perennial issues as whether we should be scientific realists and whether classifications should be hierarchical. Sections 1.4 and 1.5 provide an introduction to taxonomic pluralism and surrounding issues.

The contents of this chapter are a menu of the ontological options one faces in developing a philosophy of classification. That menu should be of intrinsic interest in itself, but it will also provide a guide through various philosophical issues in biological taxonomy. As we shall see, many of these issues are one and the same whether they appear under the guise of conceptual problems in biology or ontological issues in philosophy.

1.1 ESSENTIALISM

Well-known essentialists are found throughout the history of science and philosophy. Plato and Aristotle were essentialists. Linnaeus, Lyell, and Locke, as well as a number of other seventeenth- and eighteenth-century authors, subscribed to essentialism. And just in the last twenty-five years, Putnam and Kripke have advocated an essentialist approach to natural kinds and the semantics of natural kind terms. Essentialism, like many "isms," is not a single position, but a set of positions with varying tenets. This section will not attempt to describe each form of essentialism; there are far too many. Instead, a blueprint that captures many of the common tenets of essentialism will be offered. Then, in an effort to display some of the subtle differences among essentialists, the details of Aristotle's and Locke's forms of essentialism will be presented.

16

According to essentialism, each entity has an essential feature that makes it the type of entity that it is. That feature is an entity's real essence. The real essence of an entity occurs in all and only entities of that type, and it helps us understand why entities of that type do the sorts of things they do. For the essentialist, real essences capture the fundamental structure of the world; or to use Plato's phrase, they "carve nature at its joints." Furthermore, knowledge of such real essences is of the highest pragmatic value: if accessible, they help us identify what type of thing an entity is, and they provide a means for explaining and predicting the behavior of the world's entities. Mendeleev's periodic table is often cited as a model for essentialism. All and only the members of a particular element share a common real essence – their unique and common atomic structure. And knowledge of that structure enables us to predict and explain the behavior of instances of that element.

There is more to essentialism than real essences. The members of a kind share common necessary properties as well. Such properties are caused by the real essence of an entity and are properties that an entity must have if it has that real essence. Necessary properties are *necessary* for membership in a kind. The real essence of gold, for example, is the unique atomic structure of that element. That essence causes pieces of gold to have the necessary properties (or dispositions) of being soluble in certain types of acids, reflecting certain wavelengths of light, and having a particular range of malleability. Moreover, an entity is gold only if it has those necessary properties. The explanatory and predictive usefulness of essentialism resides in showing the real essence of a kind and what necessary properties are caused by that essence. If one knows the real essence of an entity, one can *explain* why that entity has a particular necessary property: that property is caused by that entity's essence. And if one knows the real essence and the particular circumstances of an entity, one can *predict* that it will have a certain necessary property – for example, that a chunk of gold will melt when heated to a certain temperature.

Entities also have accidental properties – properties an entity can lack without lacking its real essence. Being located on Bill Clinton's wrist is an accidental property of a particular piece of gold. That entity need not be located on Clinton's wrist to be gold; for example, through a diplomatic gesture it might come to be located on Tony Blair's wrist, yet it would still have the atomic structure that makes it gold. Despite the accidental nature of such properties, they can occur in all and only the members of a kind. In the old scholastic example, all and only humans have the property of being a featherless biped. Nevertheless, a person could lack this property and still retain the real essence of humanity – being a rational animal. When a set of accidental properties picks out all and only the members of a particular kind, that set constitutes the kind's nominal essence. Nominal essences are useful for identifying the members of a kind, but they do not satisfy the essentialist desideratum of providing classifications based on those traits that make entities the kinds of things that they are. Thus nominal essences provide provisional scientific classifications that may or may not be substantiated with the discovery of real essences.

This explanation provides a thumbnail sketch of essentialism, but it does not reveal the complexity one finds in various forms of essentialism. In order to provide a fuller picture of essentialism, I turn to Aristotle's and Locke's accounts of essentialism. Their forms of essentialism are paradigmatic in the history of philosophy, as well as the history of science. Moreover, the details of Aristotle's and Locke's accounts will be raised at important junctures in later chapters.

Perhaps no other name in the history of science is more strongly associated with essentialism than Aristotle's. But Aristotle's essentialism is much more sophisticated than the essentialism many authors attribute to him (Lennox 1987, 339). For Aristotle, each particular substance contains a nature or real essence. Indeed each particular is an instantiation of that essence. Other particulars can share that common essence – for example, there is more than one instance of animal in the world. What distinguishes particular instances of an essence are their numerically different underlying matter and how those essences are instantiated (Ayers 1981, 249; Lennox 1987, 341ff.).

Before talking about how real essences are instantiated according to Aristotle, it is important to see that Aristotle was a teleological essentialist, not a material essentialist (Lennox 1987, 340, 358–9). A material essentialist believes that real essences are physical properties. For example, the real essence of gold is its atomic structure, and the real essence of water is its molecular structure. Aristotle, on the other hand, believed that a particular's real essence is its powers to achieve certain ends. In the case of animate substances, those ends are achieving a certain kind of life. For all organisms, it is merely life itself; for animals, a life of sensation and self-movement; and for man, a life with those features as well as rationality (Ayers 1981, 251).

By conceiving of real essences as teleological functions, Aristotle allows a wide degree of variation in the possible instantiations of a real essence. If the real essence of being an animal is the possession of the powers of sensation and self-movement, then this essence can take on different instantiations, depending on the type of animal: some animals move with their fins, others with their legs, still others with their wings. Furthermore, there can be variation within a particular kind of animal: some birds use their legs to walk, some use their wings to fly, some use their wings to hobble. Which instantiation a particular animal will manifest depends on its environment (both ontogenetic and external). Aristotle offers the following example.

> Whichever birds live in swamps and eat plants have a broad beak, this sort being useful for digging and for the uprooting and cropping of their food (*PA* iv.12, 693a15–17).

> Certain of the birds are long-legged. And the cause of this is that the life of such birds is marsh-bound; and nature makes organs relative to their function, not the function relative to the organ (*PA* iv.12, 693b12–15).

All that limits variation in the physical instantiation of a real essence is an instance's ability to fulfill its essential function in its environment. Since environments can vary infinitely, so can the physical instantiations of a real essence. Perhaps Aristotle believed in the notion of supervenience – the idea that a functional property may have many physical instantiations, though no single type of physical instantiation is essential.

Aristotle not only allows variation in the visible instantiation of an essence, he also allows that some instances of an essence may not be visible at all (Sober 1980, 255ff.; Ayers 1981, 254). Some animals, for example, may have the power of self-movement but other forces prevent its visible manifestation. A donkey may have paralysis, but that is just a force counteracting the donkey's power of self-movement. Even if a donkey is born with no legs it still has the power of self-movement; it just cannot display that power through leg movements. On the face of it, the suggestion of a power that shows no outward manifestation may seem unscientific by today's standards. But often we cannot observe forces at work when their effects are masked by countering forces. I may push against a boulder with all my might, but the boulder does not move because of the countering forces of gravitation and friction.

Let us now turn to what role those essences play in classification and

Aristotle's suggested methodology for finding such essences. The topic here is confusing at first glance because Aristotle provides two different perspectives. In his *Metaphysics*, Aristotle introduces a system of classification from a logical point of view, whereas his biological works, *Historia Animalium*, *De Partibus Animalium*, and *De Generatione Animalium*, contain discussions from the standpoint of someone constructing classifications by empirical means (Atran 1990, 106ff.; Lloyd 1968, 86ff.).

The *Metaphysics* introduces the genus/species distinction and the method of dichotomous division. The real essence of a species is a combination of its genus and the *differentia* that distinguishes it from the other species of that genus. For example, the species man belongs to the genus animal, and its *differentia* is rationality. Hence the real essence of man is being a rational animal. According to the method of dichotomous division, a proper classification consists of a hierarchy of classes, each defined by the genus it belongs to and its *differentia* within that genus. Furthermore, a genus should be dichotomously divided according to which entities have a particular *differentia* and which do not. The genus of animate objects is divided into animals and vegetables according to the *differentia* of self-movement. Moving down one level in the hierarchy, the genus animal is divided into man and the lower animals according to the *differentia* of rationality (Ayers 1981, 251).

Despite Aristotle's introduction of the method of dichotomous division in his earlier works, he argues against using that method in his biological works. In *De Partibus Animalium* he suggests that dividing animals into such pairs of opposites as footed and footless splits natural groups (642b10). For example, being blooded or not is a lower division than being footed, and according to the method of dichotomous division, all blooded animals should be either footed or footless – they should not occur on both sides of a higher division. Yet all footed animals and some footless animals (for instance, snakes) are blooded. So the method of dichotomous division breaks up the natural group blooded animals.

In order to preserve natural groups, Aristotle suggests a different method for arriving at classifications in *De Partibus Animalium*. Instead of initially dividing a higher group according to a single *differentia*, divide it along many *differentiae* simultaneously (Atran 1990, 108; Lennox 1993). Once a division has been established via a number of traits, further analysis may reveal that there is a unifying difference for that group. For example, Aristotle (*PA* iii 3, 6) observed that many animals tended to have lungs, windpipes, and esophagi (though these organs occur in various forms). By analyzing the relations among these *differentiae* he discovered the difference that organizes them: all three contribute to the power of being air-cooled. He concluded that animals are divided into those that are air-cooled and those that are water-cooled (fish).

We have seen that Aristotle's essentialism is much more complex than the caricature often attributed to him. Real essences are allowed to have various instantiations, and some instantiations may even lack a visible manifestation. Furthermore, his method for discovering such essences is not one of pure logical analysis but an empirical one that involves the consideration of a number of traits.

The other paradigmatic account of essentialism I would like to discuss is John Locke's. Locke's essentialism departs from Aristotle's in important ways. One difference is how each author conceived of real essences. For Aristotle, real essences are powers or functions. Real essences are embodied in matter, but they are in no way material; they are pure form. Locke, on the other hand, did not want anything to do with such "occult" powers and functions (Ayers 1981, 254). Locke employs Boyle's corpuscularianism: all macroscopic entities are made up of solid particles clashing in the void (Ayers 1981, 250). Consequently, he held that real essences are nothing more than the various shapes, sizes, and motions of minute particles that underlie macroscopic objects. So for Locke, an oak tree is an oak because of the structure of its minute particles, not because it participates in a certain nature or form.

Locke's and Aristotle's varying notions of real essence have implications for their views on identity and change. For Aristotle, an individual entity exists only as a certain kind of thing: once it leaves its kind it is no longer the same entity (Ackrill 1981, 124). The real essence of Charles Darwin is his rationality. If he loses that essence he ceases to exist. For Aristotle, form (real essence) takes precedence over matter in individuating entities. Locke holds a different view. An individual entity is a discrete and coherent "machine" and can remain the same individual even if it changes kinds. A caterpillar can become a butterfly and still be the same individual. Locke even allows that a single entity can be a horse at one time and a cow at another (Ayers 1981,

270–1). An individual ceases to exist only when it no longer coheres as a discrete physical unit (Ayers 1981, 271). Thus for Locke identity over time depends only on the persistence of matter, whereas for Aristotle it depends entirely on persistence of form.

What of our ability to come to know the real essences of things? Although Aristotle writes at some length on the difficulties of coming to know real essences, he believes that we can know them. Locke, however, is less sanguine. For Locke, a nominal essence consists of the macroscopic properties that mark the boundaries of a kind: "the *nominal Essence* of *Gold*, is that complex *Idea* the word *Gold* stands for, let it be, for instance, a Body yellow, of a certain weight, malleable, fusible, and fixed" (1894, III, vi, 2). Ideally, the nominal essence of gold will be caused by the microstructure that constitutes the real essence of the kind gold. But how do we know that the nominal essence of the word "gold" corresponds to a particular real essence that would ontologically ground the kind gold? According to Locke, we need to inspect the microstructure of those objects we call gold. But there lies the epistemological problem. Locke again:

> Nor indeed *can we* rank, and sort *Things*, and consequently (which is the end of sorting) denominate them *by their real Essences*; because we know them not. Our Faculties carry us no farther towards the knowledge and distinction of substances, than a collection of those sensible *Ideas* which we observe in them. . . . A blind Man may as soon sort Things by their Colours, and he that has lost his Smell, as well distinguish a Lily and a Rose by their Odors, as by those internal Constitutions which he knows not (1894, III, vi, 9).

Earlier in the *Essay*, Locke regrets our lack of microscopic eyes (1894 II, xxiii, 12), for without them Locke thought we have no access to real essences. So according to Locke, we have no way of knowing whether the groups of objects picked out by our nominal essences correspond to real essences in nature.

Some might find Locke's skepticism premature. This is the stance of such contemporary philosophers as Kripke (1972), Putnam (1975), and Ayers (1981). Chemistry and physics have provided instruments that allow us access to the world of microphenomena. Furthermore, we have found that many antecedently established groups contain members who share similar microscopic properties. Has modern science then circumvented Locke's skeptical concerns over essentialism? Not entirely. Even if we had access to microphenomena, Locke thinks there would still be the problem of locating real essences. Consider his comments on the occurrence of vagueness at both the macroscopic and microscopic levels:

> There are Animals so near of kind to Birds and Beasts, that they are in the Middle between both. . . . There are some Brutes, that seem to have as much Knowledge and Reason, as some that are called Men . . . and so on till we come to the lowest and the most inorganical part of Matter, we find everywhere, that the several Species are linked together, and differ in almost insensible degrees (1894, III, vi, 12).

Even if we had access to the world of microscopic phenomena, the distinctions between microstructure might be so fine that we cannot divide them into distinct real essences.

Does this last argument indicate that Locke was a skeptic about finding real essences yet believed they existed, or did he hold that there were no real essences in nature? Ayers (1981, 260) contends that Locke did not believe in the existence of real essences but only mentioned them to establish his skeptical arguments. Kornblith (1993, Chapter 2), on the other hand, maintains that Locke was a realist when it comes to real essences and a skeptic concerning whether the science of his day could pick them out. Whatever view one takes of Locke, the question of whether vagueness undermines our ability to establish natural classifications cannot be brushed aside. We will return to that question and give it more attention in Sections 1.3 and 3.1.

Aristotle's and Locke's accounts of essentialism vary in important and interesting ways. Nevertheless they agree on the main tenets of essentialism: all and only the members of a kind share a set of traits; Those traits make entities the kinds of things they are; thus, those traits are crucial in explaining the other properties typically associated with the members of a kind. A number of philosophers and biologists have been suspicious of these tenets (including Locke himself). For example, Darwin (1859, 53, 485), and Adanson a hundred years earlier (see Hull 1965, 205), doubted whether any traits could be found among all and only the members of a species. Russell and Wittgenstein expressed similar doubts outside the biological realm (see Hacking 1991a). The inability to find traits among all and only the members of recognized natural groups is just one of the problems seen with essentialism. There are many others.[1] We will return to essentialism and evaluate its prospects in biology in Chapter 3. We now turn to two different approaches to classification.

1.2 CLUSTER ANALYSES

Just as there are different types of essentialism, there are different cluster approaches to classification. All forms of cluster analysis make two common assumptions: the members of a taxonomic group must share a *cluster* of similar traits, and those traits need not occur in all and only the members of a group. Still, cluster analyses vary: first, on the breadth of similarities desired among the members of a taxonomic group, and second, on the relationship between similarity and theory.

Wittgenstein's notion of family resemblance is the least restrictive of cluster analyses. His is a disjunctive account of classification, as opposed to the conjunctive approach offered by essentialism. Conjunctive accounts require that the members of a taxonomic unit share a conjunction of traits. According to essentialism, each trait of such a conjunction is necessary for membership and the totality of traits is sufficient for membership. Wittgenstein's disjunctive account, on the other hand, classifies a set of entities according to a disjunction of traits. Consider Wittgenstein's analysis of the group of things we call games:

> [Y]ou will not see something that is common to them *all*. . . . Look for example at board-games, with their multifarious relationships. Now pass to card-games; here you may find many correspondences with the first group, but many common features drop out, and others appear. When we pass next to ball-games, much that is common is retained, but much is lost. . . . I can think of no better expression to characterize these similarities than "family resemblances" . . . (Wittgenstein 1958, 31e–32e).

Games are classified according to a disjunction of traits, where each disjunct contains more than one trait and the traits within a disjunct can vary. Suppose capital letters represent traits of games, then a definition of game will look something like this: All games have (A and B and C) or (A and D and F) or (D and Z and Q) or . . . , and so on. The actual definition of game will be much more complicated since each disjunct will contain many more traits and the disjunction itself will contain many more disjuncts. In fact, this disjunction must be left open-ended until we have examined all the traits of each game, something no particular human is likely to do. Also notice that for Wittgenstein, each disjunct of traits is sufficient for an entity to be a game, but no disjunct is necessary.

One other point concerning Wittgenstein's cluster analysis is worth mentioning. Why does he call the relations among entities in a taxonomic unit "family resemblances"? The word "family" should not mislead one to think that the resemblances among games have a genealogical basis. Some games may have a common origin, but that is not what Wittgenstein was emphasizing. For Wittgenstein, family resemblances are *serially* related resemblances. Suppose the games in one set share traits A and B and C, the games in a second set share traits A and B and D and F, and games in a third share traits D and F and G. The games in the first two sets directly resemble one another; the games in the second and third sets directly resemble each other; but the games in first and third sets only indirectly, or serially, resemble one another.

We have seen that Wittgenstein offers a *disjunctive* approach to classification, whereas essentialism is a *conjunctive* approach where all and only the members of a kind have a conjunction of traits. Between Wittgenstein's notion of family resemblance and essentialism lie other conjunctive cluster approaches to classification. These approaches place more restraints on classification than Wittgenstein's, yet they place fewer constraints on classification than essentialism.

Consider the phenetic or numerical school of biological taxonomy. Pheneticists (for example, Sneath and Sokal 1973) recommend constructing classifications according to those groups of organisms with the most overall similarity. Taxonomic units, in other words, should be groups of entities that share a large number of common properties. Pheneticists desire classifications based on which entities share the most overall similarity for a couple of reasons. One is the belief that classifications serve as the basis for making inferences. The more common properties found among the entities of a taxonomic group, the higher is the "predictive value" (Sokal 1985, 739) of a classification. So the more overall similarity revealed by a classification, the more inferences can be made using that classification and the better the classification. The other reason that pheneticists promote classifications based on overall similarity is their explicit desire to rid classification of theoretical bias. The notion of mere similarity, they suggest, is not dictated by any theory and can be used to construct theoretically neutral classifications.

The next chapter will provide a more thorough introduction to the phenetic school of taxonomy; and later chapters will consider the feasibility of theory neutral approaches to classification. Nevertheless, we can see that pheneticists place a stronger constraint on classifications than that imposed by Wittgenstein's notion of family resemblance. Pheneticists require that the members of a taxonomic unit share a number of similar traits, whereas Wittgenstein's cluster account requires only that the members of a kind serially resemble one another. Though the cluster account of pheneticists is more restrictive than Wittgenstein's, it is still weaker than essentialism. No traits are considered either necessary or sufficient for membership in a taxonomic unit. Furthermore, phenetic classifications are not supposed to be implied by theory but are thought to be distinct from theory.

Hempel also advocates a cluster approach to classification, but one that ties theory and classification closer together. For the pheneticists, any similarity counts as a basis for clustering entities into a taxonomic unit. The more similarities the better, for they afford more inferences about those entities. Hempel, however, wants scientific classifications to do more than provide a ground for inference and prediction; scientific classifications should also help us understand the world. Thus for Hempel, only certain types of similarities are important, namely, those that provide a basis for "explanation, prediction, and general scientific understanding" (1965, 146). But how do we determine which similarities provide scientific understanding? According to Hempel, we want clusters of entities that display those similarities highlighted by scientific laws. As Hempel puts it, scientific classifications should have *systematic import*: they should systematize the knowledge contained in a field's "laws or theories" (ibid., 146–7).

Hempel's approach to classification is more constrained than either Wittgenstein's account or pheneticism, for only those clusters containing theoretically important similarities count. It is also more ontologically robust than their accounts. Classifications that group entities according to theoretically important similarities are considered more objective than those that group entities by mere similarity. For example, Hempel writes,

> the division of humans into male and female is associated, by general laws or statistical connections, with a large variety of concomitant physical, physiological, and psychological traits. It is understandable that a classification of this sort should be viewed as somehow having objective existence in nature, as "carving nature at the joints", in contradistinction to "artificial" classifications, in which the defining characteristics have few explanatory or predictive connections with other traits; as is the case,

for example, in the division of humans into those weighing less than one hundred pounds, and all others (1965, 147).

Scientific theory, in the form of lawlike generalizations, indicates which groups of entities should be recognized in scientific classifications. Moreover, theory guides us in deciding which groups occur naturally.

Hempel's cluster analysis has several of the virtues of essentialism. His taxonomic units are the basis of prediction and explanation. His cluster analysis also provides a means for discerning more or less artificial groupings. Still, Hempel's account is more modest than essentialism. It does not require that any of the similarities found among the members of a kind be necessary or sufficient for membership in that kind. Furthermore, it does not say anything about the metaphysical underpinnings of natural groups. Essentialism, on the other hand, asserts that the members of a natural group share a common essence that causes them to be members of that group.

Recently, Boyd (1990, 1991) has offered a cluster approach to classification that explicates the metaphysical structure of natural groups without invoking essentialism. Boyd believes that scientific classifications should highlight *homeostatic property cluster kinds*. Like previous versions of cluster analysis, save Wittgenstein's, the members of a homeostatic property cluster kind share a "cluster of co-occurring properties" (Boyd 1991, 141). And like Hempel's account, relations among those properties are lawlike and allow us to predict and explain the behavior of entities in such kinds. All of this we have seen before in one or more versions of cluster analysis. Boyd's original contribution is the suggestion that the co-occurrence of theoretically important similarities in natural groups is the result of underlying *causal homeostatic mechanisms* (1990, 373). Homeostatic mechanisms *cause* the properties typically found among the members of a kind. They are the ontological bedrock of natural groups. They cause the properties of kinds to be sufficiently stable that we can successfully make predictions and explanations.

Boyd believes that a number of scientifically important groups are causal homeostatic cluster kinds. Chemical elements are one example. The shared subatomic structure of all pieces of gold is the homeostatic mechanism of that group. That microstructure causes the other properties that typically occur in pieces of gold. This example might make Boyd's account sound like a variant of essentialism, but it is not.

Homeostatic cluster kinds may have necessary and sufficient properties, but they need not have such properties.

> Imperfect homeostasis is nomologically possible or actual: some thing
> may display some but not all of the properties in [a kind]; some but not
> all of the relevant underlying homeostatic mechanisms may be present
> (Boyd 1990, 373).

As another example of homeostatic cluster kinds, Boyd cites sexually reproducing species. According to Boyd, the morphological, physiological, and behavioral features that characterize the members of
such species are caused by the homeostatic mechanism of interbreeding (1991, 142). But not every member of a species must have all of a
species' typical characteristics. Similarly, some members of a species
may lack that species' homeostatic mechanism – for example, some
members of a species may be sterile and cannot interbreed.[2] Stepping
back, we see that membership in a homeostatic cluster kind is not
defined by necessary and sufficient conditions. And causal homeostatic mechanisms are not essences: the causal homeostatic mechanisms
of a kind can vary. Boyd's cluster account has many of the virtues of
essentialism yet it pulls back from attributing full-blown essences to
kinds.

1.3 THE HISTORICAL APPROACH

Cluster analysis and essentialism sort entities according to their qualitative similarities. We have seen a wide range of opinion on the nature
of those similarities, but none of those approaches places a requirement on the causal relations *among the entities* being classified. All entities sharing a certain atomic structure, according to essentialism, are
gold, independent of their causal relations to one another. On the pheneticist's and Hempel's brands of cluster analysis, entities that share an
extensive number of similar traits form a kind regardless of their causal
connections. Boyd's homeostatic approach is a little different. Boyd
talks about the causal relations *among the similarities* of the members
of a kind, but he does not emphasize nor does he require any causal
relations among the members.

The historical approach to classification reverses the roles of qualitative similarity and causal relations: causal relations are primary
and qualitative similarity is important only when it serves as evidence

28

for causal relations. Consider Darwin's suggestion for classifying organisms:

> All true classification is genealogical; that community of descent is the hidden bond which naturalists have been unconsciously seeking, and not some unknown plan of creation, or enunciation of general propositions, and the mere putting together and separating of objects more or less alike (1859, 420).

According to Darwin's suggestion, humans, for example, should be sorted into the species *Homo sapiens* not because they look alike, but because all humans form an uninterrupted and unique causal sequence of organisms.

Underlying the historical approach to classification is a metaphysical thesis. A taxonomic group consists of entities that have certain causal relations to one another. Furthermore, those relations cause the entities of a taxonomic unit to form a spatiotemporally bound entity. All humans, for example, constitute a single taxonomic unit – a species – because all humans are connected either directly or transitively by heredity relations. Those heredity relations in turn cause the species *Homo sapiens* to be a particular genealogical chunk on the tree of life. Merely being an aggregate of spatiotemporally contiguous entities is not enough for forming a taxonomic unit on the historical approach. Being parts of a unique and uninterrupted causal sequence is essential.

A question arises: Why should one classify entities according to their forming causally bound wholes? According to essentialism and most versions of cluster analysis, classifications are constructed according to qualitative similarities that can be used for explaining or predicting the behavior of a kind's members. The periodic table, for example, helps us explain why a chunk of gold placed in sulfuric acid dissolves by highlighting its atomic structure. Historically based classifications, on the other hand, help us explain an entity's properties that are due to a sequence of events by highlighting the causal path underlying those events. Consider one of Gould's (1980) examples: Why do pandas have a false thumb fashioned from the wrist's radial sesamoid bone? Because the true anatomical first digit in their carnivorous ancestors evolved to a limited use in clawing. Descendant herbivorous pandas, however, developed a sesamoid "thumb" for stripping leaves off bamboo shoots, and those pandas survive to this day. To understand why today's pandas have sesamoid thumbs, we need to trace their evolution from their carnivorous ancestors, to the introduction of sesamoid thumbs in herbivorous pandas, to the persistence of those thumbs in today's pandas. In other words, we need to know the causal sequence of events leading up to contemporary pandas which in turn requires knowing the genealogical lineage that runs from carnivorous pandas to today's pandas.

Thus far only biological examples have been used to introduce the historical approach. But that approach is not unique to biology. Other disciplines, such as geology, cosmology, human history, and linguistics, construct classifications according to historical connections. In fact, any discipline that identifies the parts of an entity over time or identifies the path of a causal process employs the historical approach to classification. Arguably the historical approach is employed in every discipline, for the historical approach merely attempts to individuate entities. Sometimes we call that activity "constructing a historical taxonomy," other times we refer to it as "individuating an entity"; but they are the same type of project. Often we call the act of individuating "classification" when the components of an entity are not obviously components of a single entity. Species, for example, consist of organisms that are causally connected by interbreeding or begetting. Yet such processes connect those organisms for only a small portion of their lives. So when we view a species, we do not see a causal connection among its organisms. Hence we tend to treat the activity of individuating a species as classifying organisms into a taxon. (More will be said about the causal connections among members of a species in the next two chapters.)

The above introduction to the historical approach to classification reveals a myriad of issues in need of further analysis. Much of this book is an examination of the historical approach as it is used in biological taxonomy. For that reason, and because philosophers have paid less attention to it than to the qualitative approaches, I will delve more deeply into some of the philosophical issues that arise in the historical approach.

Causal Connections

We have seen that the historical approach to classification identifies which objects are parts of particular historical entities. And we have seen that being a member of (or more correctly, a part of) a historical entity turns on the causal connections among a group of objects. But much more should be said about that nature of those causal connections. To begin, there are two types of causal connections found in historical entities: causal connections among parts that exist at different times, and causal connections among parts that exist at the same time. Minimally, all historical entities consist of objects that are causally connected over time. My father, myself, and my son are parts of a biological family only if we are appropriately causally related through time; we are all connected through various reproductive processes. Notice that my father and my child need not be directly connected by such processes to be parts of the same family, but serially connected. Similarly, Betsy as a two-year-old, Betsy as a twenty-year-old, and Betsy as an octogenarian are all temporal stages of a single human if those stages are appropriately related to intermediary stages by physiological and developmental processes.

These examples bring to the fore two important points about historical entities. When we ask whether a series of objects forms a historical entity we should not ask whether they form a historical entity *simpliciter*; otherwise our answer will be ambiguous. For example, when asking if my father, myself, and my son are parts of a single historical entity, are we asking if they are parts of a single family or if they are parts of a single human? Depending on which way we formulate the question, we will get different answers. To avoid ambiguous answers, we need to ask if particular objects are parts of a historical entity of a certain type. This leads to the second point. By specifying the type of historical entity under investigation we place constraints on which causal relations are appropriate for an entity. The parts of a historical entity must be causally connected in ways specific to the type of entity in question. It is important to clarify which causal connections are appropriate or we may get the wrong answer for the investigation under study. For instance, in asking whether George Bush and Toni Morrison are parts of the same family, we do not merely want to ask if they are causally connected. They do exert certain gravitational forces on one another, but that does not make them members of a single family.

Appropriate causal connections are necessary for making a historical entity a certain type of entity. But how do we know which causal connections are appropriate for a historical entity? We do so by turning to the scientific theory(ies), if there is one, that studies the type of historical entity in question. Physiology and developmental biology, for example, tell us what causal processes bind different stages of an organism. Evolutionary theory tells us which processes connect organisms into a single species. At this juncture, two caveats should be mentioned. First, the scientists that study the type of entity under investigation may disagree on which causal connections are appropriate. As we shall see in the next chapter, that is certainly true for biologists when it comes to the nature of species. Second, some historical entities are not studied by science. The causal processes that bind their parts are determined by social rules or customs. Consider the question of whether various baseball game innings are innings of a single game.

We have discussed the causal connections that bind parts that exist at different times. Let us turn to those causal connections among parts that exist at the same time.[3] A single cup is a historical entity consisting of parts that exist at a single time. For those parts to be parts of a single cup they must be appropriately causally connected. Here electrostatic forces are the appropriate causal processes that bind cup parts into a historical entity that one can drink from. Such forces, I am told, work constantly among the parts of a single cup; otherwise, the parts would fall away from one another. Not all historical entities, however, consist of spatial parts (parts existing at the same time) that have a constant causal connection. Social organizations form historical entities yet there are times when their members are not socially interacting. The causal interaction among the spatial parts of some historical entities is intermittent. The battalions of a single army do not always socially interact, nor do all the members of the Pittsburgh Pirates baseball team. Turning to biology, some species consist of populations living on distant oceanic islands; as a result, members of those populations infrequently interact with one another.

These examples raise the following question. How frequently must the spatial parts of historical entities causally interact? Again, we turn to the scientific theories or social conventions that govern the types of entities in question. How often must the populations of a species appropriately interact in order to remain parts of a single species? To answer that question we should turn to evolutionary biology and empirical studies conducted by biologists. However, we will not necessarily get a single uncontroversial answer, for as we shall see in upcoming chapters, biologists disagree widely on this issue. They disagree not only on the empirical findings of how much interaction is needed but also on the more theoretical question of what constitutes a species. Often when we turn to science to help us answer such metaphysical questions, we do so not because science has the answers, but because it contains the best understanding of the entities under investigation.

At this point, let me suggest that the frequency of interaction required by the spatial parts of historical entities lies on a continuum. At one end are those entities whose existence requires fairly constant interaction among their parts. Cup parts require constant interaction in order to remain parts of a single cup. The battalions of an army and the ants of a single colony, on the other hand, are examples of entities that lie somewhere in the middle of that continuum; their parts must interact occasionally but constant interaction is not needed. At the other end of the continuum are those historical entities whose parts need not have any relevant ongoing causal connections. The genera *Pongo, Pan, Gorilla*, and *Homo* are all parts of a single historical entity, the family Homoinidea, because they are all genealogically connected to a common and unique ancestor. However, membership in that family does not require any biological interaction among the members of those genera. Though Homoinidea is a historical entity, it is not one whose members must biologically interact.

The continuum of causal interaction required among the spatial parts of historical entities gives rise to a continuum of types of historical entities. Some historical entities have a high degree of causal cohesion among their spatial parts, some have less cohesion, and others have none. Rosenberg (1985a) and Ghiselin (1987) call all historical entities "individuals." Others reserve the word "individual" only for those historical entities whose spatial parts causally interact (Guyot 1987, Ereshefsky 1991a). I am not so interested in determining the correct usage of the word "individual" as making clear that the category "historical entity" consists of ontological variants. Some historical entities are merely historical entities – their spatial parts are not causally connected; other historical entities have a high degree of causal cohesion among their spatial parts. As we shall see in Chapter 3, these ontological distinctions are important, not the words we attach to them.

The flip side of seeing what causal connections are required among the parts of a historical entity is seeing how much variation is allowed among those parts. Both over time and at a time, the qualitative properties of a historical entity can vary greatly. Consider the tree of life on this planet. Biologists allow that life may have originated more than once on Earth, but the common assumption is that current life is descended from a single origination event (see Sober 1988, 8, for references). Consequently, all of current life on this planet and all of its ancestors are parts of a single historical entity – a genealogical tree rooted to a single inception of life. If one follows a single line of organisms through time on that tree, the variation of qualitative traits among those organisms will be quite extensive. Similarly, if one looks at all contemporary organisms (that is, all contemporary parts of the tree of life), one will find extensive variation.

One might wonder if there is a limit to how much qualitative variation can occur in a single historical entity. The parts of a historical entity can vary qualitatively from one another as long as that difference does not preclude them from being parts of that entity. What we need to ask is whether there are qualitative differences that disrupt historical entities. All organisms on the tree of life are carbon-based life forms. Is having that qualitative property necessary for being a part of that genealogical tree? That is an empirical question, not a conceptual one. Suppose, according to our best scientific theories, no non-carbon-based organism can be genealogically connected to a carbon-based organism on this planet. Then there would be a qualitative property that all the parts of that historical entity must have. The point of this example is to show that limits may exist on the qualitative variation among the parts of a historical entity. But notice that such limits are constraints placed on the tree of life so that it can be a historical entity. Thus, such constraints in no way diminish its being a historical entity.

Just as the qualitative traits of a historical entity can vary, so can the causal relations that bind those parts. For example, the spatial parts of a building may be connected by different causal processes: some parts may be connected by cement, others may be joined by metal rivets, others stay together due to the force of gravity, and so on. Causal links between the temporal parts of a single historical entity can vary as well. The lizard genus *Cnemidophorous* consists of species of organisms that reproduce sexually and species of organisms that reproduce asexually (Kitcher 1984a). The eggs of the females in the sexual species must be fertilized by males while the eggs of the females in the asexual species are self-fertilized. So two different reproductive processes diachronically connect members of the genus *Cnemidophorous*. In both of these examples, the causal links among the parts of a historical entity are allowed to vary, but that variation is constrained by how much variation is permissible for the type of entity in question.

Vague Boundaries

In Section 1.1 we saw Locke's concern that vagueness may undermine an essentialist approach to classification. Every approach to classification must wrestle with the problem of vagueness, including the historical approach. Consider how vagueness may arise in a historical classification. Suppose a small population of tortoises lives on a small peninsula stretching into the ocean. Subsequently, the ocean breaks through the peninsula, causing the tortoises to live on an offshore island, geographically isolating them from the rest of their species. Over time, the environment on the island comes to differ from that of the mainland, and the mating habits of the island tortoises start to vary from those of the mainland tortoises. Here's the question. Do the island tortoises and the mainland tortoises belong to the same species or different species? According to the most widely held definition of species, Mayr's Biological Species Concept (1970), the island population is a distinct species only if its organisms are "reproductively isolated" from the mainland tortoises. That is, the island population constitutes a different species only when its members have biological mechanisms that prevent them from interbreeding and producing fertile offspring with the mainland tortoises. But when does that occur? When such a mechanism first arises within a member of the island population? When the majority of organisms have that trait? When all of them have that trait?

None of these precise answers will do. One new organism does not make a new species, because biologists view species as distinct populations that have succeeded for multiple generations. Requiring that a majority of organisms have a common reproductive isolating mechanism won't do either: the population may consist of only five organisms, which subsequently die; hence no distinct population has succeeded for any length of time. Finally, requiring that all of the organisms of a species have reproductive isolating mechanisms is too stringent because hybridization between two species does not necessary prevent those species from being distinct. Species, according to the Biological Species Concept, are distinct and successful populations that are protected (genetically) by reproductive isolating mechanisms. How long they must be successful is vague, as is the exact number of organisms that must have reproductive isolating mechanisms. (A more thorough introduction to the Biological Species Concept and its associated account of speciation will be given in the next chapter.)

I bring up this example to show that vagueness can arise in historical classifications as well as in classifications based on essentialism or cluster analyses. Does such vagueness show that the historical approach, or the Biological Species Concept, is problematic? I do not think so. The world may be an inherently vague place, where the processes that transform an entity into a different type of entity take time. During such processes the world provides no determinate answer to the question of that entity's type. Nevertheless, the existence of such transformation processes does not imply that there are no clearly distinct entities outside of those processes. Species may be an example: lots of distinct species exist, but by the same token, speciation causes a number of borderline cases. What mixture of vague and nonvague cases we should tolerate in a mode of classification is a pragmatic question. Certainly the sort of situation that Locke fears, where vagueness is so rampant that no classification can be constructed, is not a fruitful one. But on the other hand, demanding that an approach to classification give a precise answer in every taxonomic case is overly idealistic.

Identity and Individuation

The study of how to individuate historical entities has led to some notoriously intransigent problems in philosophy and biology (see Janzen 1977, Wiggins 1980, Splitter 1988, and Horvath 1997). To get a feel for these problems, let us go through a series of cases. Hydrozoa (corals, polyps, etc.) reproduce asexually when a parental organism buds off a part of itself. Parent and offspring are genetically the same, but they are physiologically independent and lead distinct lives. Suppose a parental organism buds off 10 percent from the left side of its original body. Both the resulting bud and the remaining 90 percent of the original body go their separate ways and lead distinct and successful lives. In such a case, it seems fairly intuitive that the remaining 90 percent of the original body *is* the parent. In other words, the budding does not destroy the parent.

An interesting observation concerning how we individuate historical entities arises when one compares the hydra case to another. With the hydra, 10 percent of the left side of the original body buds off. The remaining right side is the parental organism. But now suppose that 10 percent of the right side of the original body buds off. Then intuitions switch concerning which side is the parental organism. The remaining

90 percent of the left side of the original body, it seems, is the parental organism. The point of this example is to illustrate that the identification of the parental organism is correlated with being the *larger* part. Nothing distinctive about the internal physiology of the hydra in these cases bears on which side will be the continuing organism. The only difference in these two cases (besides on which side the budding occurred) is which side is larger. Mere size seems to affect how we individuate historical entities.

Let me illustrate another oddity of individuating historical entities. Some philosophers (for example, Wiggins 1980, 71; Splitter 1988, 337–8; and Kitcher 1989, 200) hold the principle that the individuation of a historical entity should depend solely on its intrinsic properties and not on any events external or subsequent to it. I would like to call this principle into question (Horvath [1997] does the same). Consider a variation on the above hydra example. Suppose a hydra splits into two parts, one containing 60 percent of the original body, the other containing 40 percent. Furthermore, suppose the 60 percent portion is immediately destroyed after fission. In such a case we would say that the remaining 40 percent portion is the original organism. On the other hand, suppose that the 60 percent portion of the original body is not destroyed. Then, using consideration of size, we treat the larger portion as the original organism. In other words, whether the smaller portion is the original organism depends on what happens to the larger piece: if the larger piece lives, we consider it and not the smaller piece the original hydra; but if the larger piece dies, we consider the smaller piece the original hydra. This example illustrates that considerations *external* to the smaller piece affect whether we classify it as a later stage of the original organism or as a separate organism. Contrary to the principle cited above, external considerations seem to affect how we individuate entities.

If the last example is not convincing, consider a case in which the original hydra is split in half – a case of symmetric fission. Call the original body A and the subsequent bodies B and C (see Figure 1.1). Suppose C is destroyed right after fission. Then we would consider A and B stages of the same organism. Suppose, on the other hand, that C is not destroyed. Then the question of what happens to the original hydra is more confusing. Here are some options:

(1) A persists as B or C, but not both.
(2) A and B and C are all parts of the original organism.

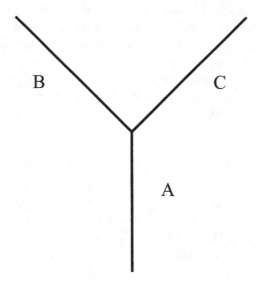

Figure 1.1   A case of symmetric fission. A is the original body. B and C are the resultant bodies.

(3) A ceases to exist after fission and is replaced by two different organisms, B and C.

Option (1) is problematic because there is no reason to think A persists as B (and not C) rather than as C (and not B). B and C have the same genetic and physiological content and the same type of ancestral connection to A. In other words, nothing distinguishes B (or C) as a better candidate for being part of the same organism as A. Option (2) is also problematic. B and C lead independent and distinct lives and we have no reason to think that B and C will later fuse into a single organism. So B and C are not parts of a single functional organism. If B and C are not parts of the same organism, then A, B, and C cannot be parts of the same organism. What about option (3)? This may be the best option. B and C lead distinct lives, and the life before, A's, no longer exists. Still, A, B, and C do belong to a common entity that began with A – that lineage or family consisting of A, B, and C. So option (3) does have some appeal: it satisfies the intuition that an entity exists before and after the splitting of A (the lineage consisting of A, B, and C), and it makes no problematic assumptions about the individuation of organisms.

One reason for introducing this case is to cast doubt on the principle that only intrinsic properties affect the individuation of an entity. We are interested in knowing whether A and B are stages of the same organism. If the case of symmetric fission in which C is not destroyed seems more problematic than the case of symmetric fission in which C is destroyed, then considerations external to A and B are affecting our decision concerning A and B. For the only difference in these two cases is what happens to C, and what happens to C does not causally affect A or B. (What happens to C does not affect A, unless one allows backward causation; and what happens to C does not physiologically affect B.) In the end, these examples cast doubt on the principle that only the intrinsic properties of an entity should affect how we individuate it. External considerations play a role as well. Why are such external considerations important? Perhaps because when we individuate an entity we individuate it against a background of other entities. In particular, whenever we ask if X is a part of an entity of a certain type, we need to check whether X might be a part of any other entities of that type.[4]

This section provides an introduction to the breadth and complexity of issues surrounding the historical approach to classification. We will return to many of these issues in later chapters. But before leaving this chapter, one other aspect of constructing classifications should be discussed. We have seen three options for constructing classifications: essentialism, cluster analysis, and the historical approach. Two more options for constructing classifications, which crisscross the above ones, are taxonomic monism and taxonomic pluralism.

1.4 TAXONOMIC MONISM AND PLURALISM

On the surface, the contrast between taxonomic monism and pluralism may seem fairly simple. Monists desire a single preferred classification of a discipline's entities. Pluralists allow a number of equally acceptable classifications of those entities. However, when one looks more closely at various positions of monism and taxonomic pluralism, the distinction gets more complicated. The task of this section is twofold. The first is to distinguish various types of taxonomic monism and taxonomic pluralism. The second is to examine the connection between taxonomic pluralism and such positions as relativism and scientific realism. The next section explores the relation between taxonomic pluralism and the desire for hierarchical classifications.

In introducing various types of taxonomic monism and pluralism, it is useful to employ two distinctions. The first is just the distinction between taxonomic monists and taxonomic pluralists given earlier: taxonomic monists desire a single preferred classification; taxonomic pluralists allow for a number of acceptable classifications. The other distinction is among metaphysical monists, metaphysical pluralists, and metaphysical agnostics. Plato's phrase "carving nature at its joints" illustrates the position of the metaphysical monist: the world is naturally divided in a single way. The metaphysical pluralist, on the other hand, believes that there are numerous divisions in nature. The metaphysically agnostic is noncommittal on such metaphysical issues. The first distinction (taxonomic monism and pluralism) concerns a preference for one or more classifications, whereas the second (metaphysical monism, pluralism, agnosticism) concerns our beliefs about the structure of the world. These two distinctions allow for a wide variety of positions conerning taxonomy, each having well-respected proponents.

The *classic* monist is both a taxonomic monist and a metaphysical monist. He maintains that we should strive for scientific classifications that accurately represent the single correct way the world is carved. Such essentialists as Plato, Aristotle, Linnaeus, Locke, Putnam, and Kripke are classic monists: they desire a single classification according to the fundamental structure of the world. (Whether they are pessimistic or optimistic about achieving that goal is another matter. Recall that Locke was pessimistic.) A classic monist need not be an essentialist. Ghiselin (1987) is a classic monist who adopts the historical approach to biological taxonomy. Ghiselin believes that nature divides the tree of life into a single nonoverlapping set of species lineages. (By nonoverlapping I mean no organism belongs to more than one species.) Furthermore, Ghiselin believes that the role of biological taxonomy is to develop a single classification of the organic world according to that single natural division.

As one might suspect, not all taxonomic monists are metaphysical monists. Hennig (1966), father of the cladistic school of biological taxonomy, is a case in point:

> Each organism may be conceived as a member of the totality of all organisms in a great variety of ways, depending on whether this totality is investigated as a living community, as a community of descent, as the bearer of physiological characters of life, as a chronologically differentiated unit, or in still other ways (Hennig 1966, 5).

> It is therefore possible to investigate the relations between different systems, which in themselves are completely and basically equally justified and equally necessary. This is done most usefully by choosing one system as the general reference system with which all others are compared (Hennig 1966, 7).

Hennig believes that the biological world consists of a number of fundamental divisions, yet for pragmatic purposes biologists should construct a single "general reference" classification. Thus Hennig is a taxonomic monist even though he believes in metaphysical pluralism. But why should we prefer a single classification of a multifarious world? Because, as Hull (1987, 1989) argues, taxonomic monism avoids ambiguity, promotes communication, and helps the advancement of science by funneling resources to the most promising project (see Section 4.4). Hennig, by the way, does not think that the choice of a preferred classification is arbitrary. For Hennig, reasons exist for promoting one type of classification over all others, and that classification is one based solely on phylogeny (see Section 2.3).

It is worth highlighting the different types of fundamental divisions that Hennig believes give rise to metaphysical pluralism in the organic world. One type of division that Hennig is *not* highlighting is that of hierarchical, or part-whole, relations. My dog Nellie is part of the species *Canis familaris*, which is part of the genus *Canis*, which is part of the family Canidae. But my dog's being a part of various genealogical entities that have different degrees of inclusiveness does not constitute a form of pluralism that distinguishes pluralism from monism. Classic monists allow that an organism can belong to multiple taxonomic groups, so long as those groups form a hierarchy of more inclusive groups. Aristotle, for example, allowed that my dog belongs to a hierarchy of kinds, and Ghiselin allows that my dog belongs to a series of genealogical lineages having different degrees of inclusiveness. What monists will not allow, and what makes for interesting forms of pluralism, is for an organism to belong to multiple groups yet those groups cross-classify the organic world. This is the sort of metaphysical pluralism Hennig highlights in the quotation above.

For Hennig, an organism belongs to "a living community" and "a community of descent," yet these communities are not hierarchically related to one another because they cross-classify the world's organisms. An insect can belong to both an ecological community and a genealogical species such that not all the members of one are members of the other. Just think of an ecological community consisting of organisms from multiple species. Similarly, if one grants that there are different types of species – for example, one based on "community of descent" and another based on "physiological characters" – then an organism might belong to two different types of species, yet those species do not contain identical members. (Empirical examples of this situation are offered in Chapter 4.) For Hennig, the organic world contains different taxonomic units that cross-classify the world's organisms. In this way he is a metaphysical pluralist, though for pragmatic reasons he is a taxonomic monist.

We have seen that some taxonomic monists are metaphysical monists, what I have called "classic monists," and some taxonomic monists are metaphysical pluralists. Let us now turn to taxonomic pluralists. One form of taxonomic pluralism that bridges the gap between monism and pluralism is a "consilence" version of taxonomic pluralism. Whewell (1840) and Ruse (1987) subscribe to this position (see Ruse's [1976] discussion of Whewell). They believe that we construct various classifications from different theoretic perspectives. In ideal circumstances those classifications coincide. When they do, that consilence provides strong evidence that we have arrived at the ontologically correct classification of the entities in question. The use of various taxonomic approaches, in other words, serves as a means for locating the fundamental divisions in nature. As an example, Ruse notes that we have various ways of classifying organisms into species – one according to genealogy, another according to genetic similarity, a third according to ecological parameters, and so on. These various ways of sorting organisms, according to Ruse (1987), give rise to the same classifications. Here taxonomic pluralism is supposed to confirm the hypothesis of metaphysical monism: different approaches to classification, each highlighting a different set of significant parameters, result in the same classifications.

Other taxonomic pluralists are either metaphysical pluralists or metaphysical agnosticists. The latter hold that for epistemological reasons we are stuck with a plurality of classifications, yet they remain agnostic about whether the world itself is monistic or pluralistic. A wide range of writers fall into the camp of epistemologically motivated pluralists. At one end are such social-constructionists, relativists, and anarchists as Latour and Woolgar (1986), Price (1992), and Feyerabend (1978). Roughly, they hold that any preference we have for one classification over another is simply a matter of taste or social convention. No objective or rational means exist for preferring *any* classification over another. Each taxonomy is epistemologically equivalent to any other, whether it be constructed by a biologist at Harvard's Museum of Comparative Zoology or a researcher at San Diego's Center for Creation Science.

Not every epistemologically motivated taxonomic pluralist is of that stripe. Some allow a plurality of equally acceptable classifications but limit which classifications should be accepted. Stanford (1995), for example, believes that we should adopt a variety of species concepts. But he does not believe that all species concepts are worthy of acceptance. According to Stanford, biologists should adopt only those species concepts that reflect our current and best biological theories. Though Stanford is a taxonomic pluralist who limits the number of acceptable classifications, he is a skeptic concerning whether our classifications are correct representations of the organic world. Classifications, for him, are the result of our theoretical interests – nothing more and nothing less. The organic world might be monistic or it might be pluralistic. We do not know and probably will never know. So, although Stanford is a taxonomic pluralist, he is a metaphysical agnostic.

Finally I turn to those taxonomic pluralists who are metaphysical pluralists. Such pluralists believe that the world is carved in multiple ways, and the most fruitful way to represent that world is with a plurality of classifications. Proponents of this brand of pluralism include Kitcher (1984a), Dupré (1993), and Ereshefsky (1992b). Here is an example of how metaphysically motivated taxonomic pluralism may arise in biology. (Chapter 4 provides a thorough argument for pluralism in biological taxonomy.) Consider the genealogical tree of life discussed earlier. Suppose the forces of evolution segment that tree into different types of basal lineages. Some lineages owe their unity to the process of interbreeding, others to ecological forces, and still others to genetic homeostasis. Moreover, some organisms belong to more than one type of lineage, and those lineages contain different members. For instance, an organism may belong to both an interbreeding species and an ecological species, yet these taxa are not fully coextensive. If that is what our best scientific theory about the biological world tells us, then the metaphysically motivated taxonomic pluralist suggests accepting a plurality of classifications that crisscross the organic world.

Such pluralism may lurk in nonhistorical approaches to classification as well. Consider an example from chemistry (Khalidi 1993, 107–8). Chemists divide substances according to unique atomic and molecular structures: nickel, iron, water, alcohol, and so forth. They also divide chemical substances according to various phases of matter: solid, liquid, and gas. These different theoretically important properties provide different classifications. Furthermore, these properties criss-cross the chemical world, resulting in a plurality of classifications that do not coincide. For instance, the molecular classification separates water and alcohol, yet ignores the difference between liquid water and gaseous water. The phase classification, on the other hand, groups liquid water and liquid alcohol together and distinguishes them from the group containing gaseous water and gaseous alcohol. Add a chemical classification based on magnetic properties (another set of theoretically significant properties) and we get a third, noncoinciding classification. Unless one is an essentialist, none of these classifications is the fundamental one for chemical kinds. Metaphysically, they are on a par.

We have seen that taxonomic pluralism may occur in both historically based and qualitatively based classifications. What about the possibility of a single discipline classifying entities according to a plurality of taxonomic methods? Such an example may be biology. Consider two ways that biologists sort organisms into significant divisions. When classifying organisms and populations using the Linnaean categories, biologists typically group those entities according to genealogy. The result is a classification of species and higher taxa consisting of historical entities. On the other hand, biologists recognize important qualitatively based distinctions that cross-classify those lineages, such as asexual organisms, host organisms, and parasite organisms. Reptilia, for example, is a historical entity consisting of both sexual and asexual organisms. But not all asexual organisms are reptiles – lots of organisms in the taxon Insecta are asexual as well (furthermore, Insecta contains both sexual and asexual organisms). So a division of organisms based on sexual similarities divides the organic world differently from one based on genealogical units.

At the beginning of this chapter, I asked if more than one general approach to classification might be appropriate for a single discipline. The above example from biology suggests that some disciplines contain both historical and qualitative classifications. Neither a qualitatively based approach nor a historically based approach is the unique correct taxonomic approach in biology, for both capture different ways the organic world is naturally carved. Thus metaphysical pluralism not only motivates taxonomic pluralism within a discipline, it also motivates methodological pluralism concerning taxonomy. In other words, it brings to the fore the need for more than one general approach to taxonomy. This suggestion for methodological pluralism concerning taxonomy echoes a suggestion made by Gould (1986), Richards (1992), and many others. They argue that different disciplines require different modes of classification; physics, for example, uses qualitative classifications, while evolutionary biology employs historical classifications. The suggestion being made here is slightly different. A *single* discipline may need to employ more than one general approach to classification. We will return to the issue of methodological pluralism in taxonomy in Section 4.4 and Chapter 5.

The metaphysically motivated pluralism outlined here has implications for two central issues in the philosophy of classification. First, it shows that taxonomic pluralism is not necessarily a relativistic approach to science. A number of people quickly associate taxonomic pluralism with the view that we should condone all suggested classifications because we lack an objective means for choosing among them. I have tried to distinguish that kind of taxonomic pluralism from a metaphysically motivated one. As indicated above, perhaps the world itself, and not our inability to make reliable contact with it, gives us reason to adopt a pluralistic stance toward taxonomy.

Second, I have tried to decouple taxonomic pluralism from an anti-realist approach to science. Roughly, anti-realists hold that we should refrain from thinking that our best theories provide, or will eventually provide, correct representations of the world. Realists, on the other hand, believe that as science progresses, our best theories will provide better and better representations of the world. Taxonomic pluralism is often coupled with antirealism on the assumption that realism and taxonomic pluralism are incompatible. I would like to suggest that there is no a priori reason that realism and taxonomic pluralism cannot co-exist. Perhaps the organic world is multifarious, and perhaps biological taxonomy has properly captured that diversity with a plurality of classifications. Notice also that taxonomic monism is not automatically a realist position. One can certainly be a taxonomic monist and an anti-realist. Instrumentalists often chose a particular taxonomic approach for pragmatic reasons but remain either agnostic or pessimistic about that approach's ability to represent the world's features accurately.

1.5 HIERARCHICAL CLASSIFICATION

The prospect of taxonomic pluralism raises the question of whether classifications should be hierarchical. At first glance taxonomic pluralism seems inconsistent with hierarchical classifications: taxonomic pluralism allows that an entity can be sorted into two taxonomic units where one unit does not fully contain the other; yet hierarchical classifications cannot allow such relations among taxonomic units. Despite initial impressions, taxonomic pluralism and the desire for hierarchies are consistent.

Let us start by spelling out what makes a classification hierarchical. A hierarchical classification consists of series of nested taxonomic units. Each unit is either completely subsumed by a higher taxonomic unit or is the highest taxonomic unit in the series. Moreover, no two units at the same level of inclusiveness can partially overlap: they must belong to different hierarchies. Consider an example from biology. An organism is sorted into a series of nested taxonomic units where each unit's level of inclusiveness is designated by a Linnaean rank. My dog Nellie belongs to the species *Canis familaris*. That species and the species wolf, coyote, and jackal comprise the genus *Canis*. *Canis* and the genus for foxes make up the family Canidae, which belongs to the order Carnivora, and so on up the Linnaean hierarchy. Nellie belongs to more than one taxon, but each taxon except the highest is subsumed by another taxon. Furthermore, Nellie belongs to no more than one taxon of the same taxonomic rank.

Hierarchical classification is the preferred mode of classification for contemporary biologists, as well as prominent essentialists in the history of biology (for example, Aristotle and Linnaeus). Why the preference for hierarchies? For the essentialist (in particular, Aristotle in the *Metaphysics*), the preference for hierarchical classification is the result of a priori beliefs concerning the nature of things. Each entity belongs to a single least inclusive kind – what I will call a fundamental kind. By knowing an entity's fundamental kind, we know its real essence and we can explain and predict its other necessary properties. This notion of fundamental kind is not violated by an entity's belonging to more than one kind, so long as all those kinds are hierarchically arranged. From an Aristotelian perspective, Bill Clinton belongs to the kinds "human," "animal," and "animate object," but these are hierarchically arranged such that human is the fundamental kind. What Aristotelian essentialism cannot allow is an entity belonging to more than one kind when those kinds are not hierarchically arranged. Otherwise, we will not know what that entity is, let alone be able to explain its nature. Suppose an entity belongs to the kind "water" and the kind "liquid." Not all water is liquid and not all liquid is water; as a result, neither kind is completely subsumed by the other. According to essentialism, only one of these two kinds can be fundamental, and membership in the other is merely accidental.

Contemporary biologists prefer hierarchical classifications as well, but for very different reasons. The standard framework for biological classification is the Linnaean hierarchy. Linnaeus developed his system of classification according to Aristotelian essentialism, not because he observed essences in nature (he admits he rarely did), but because he believed that is how God created nature (see Chapter 6). Biologists still use the Linnaean hierarchy despite its essentialist and creationist roots. But in doing so, they no longer cite essentialism as justifying their use of the Linnaean hierarchy. Contemporary biologists cite empirical, not a priori, reasons for preferring hierarchical classifications. Since Darwin, most biologists believe that hierarchical classifications represent the genealogical inclusiveness of taxa found in nature. Species are the smallest genealogical units. Genera consist of multiple species stemming from a common ancestor (though in some cases a genus may contain only one species – see Chapter 6). Genera are genealogical parts of families, and so on up the Linnaean hierarchy. Contemporary biologists (for example, Simpson 1961, 15; Hennig 1966, 20) cite this empirical fact – that nature consists of a series of inclusive genealogical entities – in justifying their preference for hierarchical classifications, not reasoning about how nature ought to be.

Nature could have produced taxonomic units that are not inclusive. That is an empirical possibility. The history of biology is strewn with nonhierarchical accounts of the organic world (Hennig 1966, 15). Lamark, for example, believed that life is ordered according to levels of perfection – what was then known as the *Scala Naturae* (Mayr 1982, 201). At the bottom of that scale are algae, mosses, and fungi; further up are plants, and at the top is the allegedly most perfect organism, man. The corresponding taxonomy is a series of taxa where no taxon is subsumed by another.

One not need turn to the history of biology to see that the hierarchical pattern of life is not a logical necessity. A nonhierarchical pattern is compatible with the existence of evolution. Entities can be historically connected (due to inheritance) and change can occur without there being hierarchical relations. Just suppose life occurred on this planet without any speciation events. Or consider a less radical possibility. Suppose speciation takes place in two ways: the splitting of a parental species into two species; and the merging of two parental species into a new and distinct species. If the latter were to occur, then a lower taxon might belong to two higher taxa where neither is fully subsumed by the other. For example, a species might be the result of two parental species merging, and those two species might belong to two different families. Consequently, the resultant species belongs to two families. The point here is not to argue for the plausibility of such examples in nature but merely to show that whether or not life is hierarchically arranged is a contingent matter.[5]

We have seen that the essentialist's preference for hierarchy is grounded in a priori reasoning, while contemporary biologists prefer hierarchies for empirical reasons. Let us now ask whether the desire for hierachical classifications is compatible with pluralism. Taxonomic pluralism allows that an entity can belong to two taxonomic units even when one unit is not completely subsumed by the other. Yet this situation violates the fundamental tenet of hierarchies: an entity can belong to two taxonomic units only if one fully encompasses the other. There is, however, a way around this problem. An entity can belong to two taxonomic units that do not completely overlap if those units are in different hierarchies. Hence taxonomic pluralism is compatible with the preference for hierarchies so long as an entity can belong to multiple taxonomies. But is that possible?

For Aristotelian essentialists it is not. An entity has only one fundamental nature and its inclusion in other hierarchies is merely accidental. But what if we are using a different mode of classification – for example, the historical approach? Does the historical approach automatically exclude the possibility of an entity's belonging equally to two different hierarchical classifications? To make this a little more concrete, recall the example of biological taxonomic pluralism given in the previous section. Suppose evolutionary processes divide the tree of life into different types of species such that an organism can belong to two different types of species at once. For example, suppose an organism belongs to a species bound by interbreeding as well as a species bound merely by genealogy. Furthermore, suppose that neither species subsumes the other. In such a case, can we say that an organism belongs equally to two species, that its membership in one is no more fundamental than in the other? What stands in the way of such a claim?

We should recognize that the question of whether evolutionary processes crisscross the tree of life is an empirical one. That question will be addressed later. Here I merely want to remove any conceptual roadblocks to taxonomic pluralism. At first glance it may seem odd to say that an organism can equally belong to two different species, that its nature, in some sense, is multifarious. This goes against the essentialist assumption that each entity has a single fundamental nature. The metaphysically motivated taxonomic pluralist, however, wants to contend that the question of an entity having a multifarious nature, intuitions notwithstanding, is an empirical one. Moreover, if one adopts approaches to classification for empirical reasons, as contemporary biologists do in preferring hierarchical and historical classifications, then the question of taxonomic pluralism should be empirically determined as well. Otherwise an inconsistency in one's methodology for choosing taxonomic approaches occurs: the preference for hierarchies and historical classifications is justified by empirical information, whereas the preference for taxonomic monism is based on a priori reasoning.

The upshot is that if one believes approaches to classification should be chosen for empirical reasons, then the acceptance (or rejection) of taxonomic pluralism should be treated no differently. The question of whether taxonomic pluralism has an empirical basis in any particular discipline remains. Chapter 4 will review the evidence for taxonomic pluralism in biology, but first, some further preliminary matters must be discussed. We have not yet had a proper introduction to the general schools and concepts of biological taxonomy. That is the task of the next chapter.

2

A Primer of Biological Taxonomy

The previous chapter provided an introduction to the philosophy of classification. In upcoming chapters we examine how various issues in the philosophy of classification arise in biological taxonomy. But first we need a proper introduction to biological taxonomy. That is the job of this chapter.

Before the schools of biological taxonomy are discussed, some terminology should be clarified. This book concerns itself with philosophical problems in biological taxonomy, but what is the difference between "taxonomy" and "classification"? And how are these distinguished from "systematics"? Let us start with the first term. According to Mayr (1969, 2), "[t]axonomy is the theory and practice of classifying organisms." A school of biological taxonomy offers principles and techniques for constructing biological classifications. Those methods tell us how to classify organisms into taxa and how to classify taxa into more inclusive taxa (species into families, for instance). The term "classification" refers to the product of taxonomy. Using the methods of a particular school of taxonomy, biologists construct classifications of the organic world.

Systematics is a bit different. In this book, I follow Simpson's (1961, 7) description of biological systematics: "Systematics is the scientific study of the kinds and diversity of organisms and of any and all relationships among them." Biological systematics does not provide methods for constructing classifications (that is the job of biological taxonomy); instead it studies how organisms and taxa are related in the natural world. Nevertheless, many biologists and philosophers hold that the results of biological systematics – what it says about the relations among organisms and taxa – should influence the principles one adopts in a school of taxonomy. Ontological considerations, in other words, should influence how classifications are constructed. As we shall

50

see, the idea that ontological considerations should influence how we construct classifications is a guiding principle in this book.

What then are the general schools of biological taxonomy? Four featured prominently in the twentieth century: evolutionary taxonomy, pheneticism, process cladism, and pattern cladism. Evolutionary taxonomy was introduced in the first half of the century and is a product of the synthetic theory of evolution. Pheneticism arose in the 1950s and 1960s, in large part, as a reaction to evolutionary taxonomy. Cladism was introduced in the 1950s and became popular among English-speaking biologists in the late 1960s. In the late 1970s, cladism divided into two competing branches: process cladism and pattern cladism. In this chapter, we will look at all four schools. Sections 2.1 and 2.2 introduce the motivations and principles of evolutionary taxonomy and pheneticism. Sections 2.3. and 2.4 do the same for the process cladism and pattern cladism.

The common aim of these schools is to provide classifications of the organic world by providing principles for sorting organisms into taxa and taxa into more inclusive taxa. One type of taxon has received special attention: species. While there are four contemporary schools of biological taxonomy, there are well over a dozen recent accounts of what makes a taxon a species. Biologists refer to these accounts as "species concepts." Section 2.5 introduces seven prominent species concepts found in the current literature.

The debate among biologists concerning general schools and species concepts has been extensive and at times rancorous. To give a proper account of these debates, or even significant portions of them, would require a book-length treatment. A handful of texts compare the positive and negative attributes of various schools (see, for example, Ridley [1986], Hull [1988], and Panchen [1992]). The intent of this chapter, however, is not to provide an introduction to these debates, but to provide (as much as it is possible) a neutral introduction to the prominent schools and species concepts in biological taxonomy. Such an introduction will provide vital background information for the philosophical issues explored in subsequent chapters.

2.1 EVOLUTIONARY TAXONOMY

The classic texts of evolutionary taxonomy are Simpson's (1961) *Principles of Animal Taxonomy* and Mayr's (1969) *Principles of Sys-*

tematic Zoology. Both authors were instrumental in forging the synthetic theory of evolution. The "modern synthesis" or the "evolutionary synthesis," as it is often called, refers to the integration of Mendelian genetics with Darwinism. When Mendel's work was rediscovered in 1900, it was thought to contradict the notion of gradual evolution offered by Darwin. DeVries and Bateson argued that Mendelism, with its notion of mutation, implied that evolution was discontinuous rather than gradual. In the 1930s, Fisher, Wright, and Haldane set about showing that Mendelian genetics and Darwinian gradualism did not contradict one another. The result of their efforts was a set of mathematical models demonstrating how Mendelism provides the proper genetic basis for gradual evolution. A remaining task was to apply the insights of the synthetic theory of evolution to biological taxonomy. That job was undertaken by Mayr and Simpson, as well as Huxley, Stebbins, and Dobzhansky. Their efforts resulted in the taxonomic school called "evolutionary taxonomy." Its guiding principle is to provide classifications that reflect the insights of the synthetic theory.[1]

According to the synthetic theory of evolution, new taxa arise through two processes: cladogenesis and anagenesis. Mayr is explicit concerning the relation between these two processes and evolutionary taxonomy. He writes, "evolutionary taxonomists . . . aim to construct classifications that reflect both of the two major evolutionary processes, branching and divergence (cladogenesis and anagenesis)" (Mayr (1981[1994], 290). In cladogenesis, a species is split in two (Figure 2.1a). Consider the case of a population that becomes geographically isolated from the rest of its species. If that population is exposed to different selection forces and exchanges no genes with the rest of the species, it may undergo a genetic revolution. During that revolution the members of the population may acquire mechanisms that prevent them from successfully interbreeding with the members in the parental species. In such a case, speciation occurs through the splitting of a single genealogical lineage.

Speciation through anagenesis is different. In anagenesis, a new species can evolve by the gradual changing of a single lineage (Figure 2.1b). However, not any change gives rise to a new taxon; the change required for speciation must be significant. One sort of change evolutionary taxonomists consider significant occurs when a lineage enters a new environment and acquires a radically new set of adaptations. The

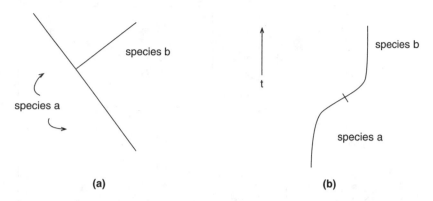

Figure 2.1   (a) A case of speciation through cladogenesis. A population of species A becomes geographically isolated from the rest of the species. It undergoes a genetic revolution and becomes a distinct species, B. (b) A case of speciation through anagenesis. Species A gradually evolves until it becomes a new species, B.

resultant taxon, according to evolutionary taxonomists, occupies a new adaptive zone. We will discuss the nature of adaptive zones at the end of this section and in Section 2.5.

Cladogenesis and anagenesis give rise to two different types of taxa, so the job of evolutionary taxonomists is to find and classify such taxa. Taxonomists refer to these taxa as "monophyletic taxa" and "paraphyletic taxa." A monophyletic taxon contains an ancestor and all and only its descendants.[2] In Figure 2.2, the taxon consisting of D, E, and B is a monophyletic taxon; so is the taxon containing H, I, and G, and the taxon containing H, I, G, F, and C. Paraphyletic taxa, on the other hand, contain an ancestor and some, but not all of its descendants. A classic example of a paraphyletic taxon is Reptilia. That taxon contains lizards, snakes, and crocodiles, but does not contain birds (Figure 2.3). The difference between Reptilia and birds, argue evolutionary taxonomists, is much more significant than the difference among lizards, snakes, and crocodiles. Hence, evolutionary taxonomists place lizards, snakes, and crocodiles in one taxon and birds in another.

Having seen the difference between monophyletic and paraphyletic taxa, we can now relate them to the processes of cladogenesis and anagenesis. In cladogenesis, new lineages evolve through the splitting of an ancestor. The result is a new monophyletic group, that is, a group containing an ancestor and all and only its descendants. In Figure 2.2,

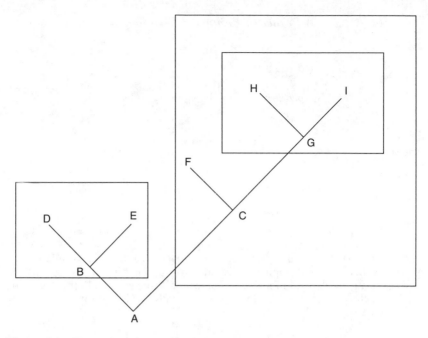

Figure 2.2   Examples of monophyletic taxa. The taxon consisting of D, E, and B is a monophyletic taxon; so is the taxon containing H, I, and G, and the taxon containing H, I, G, F, and C. Each contains an ancestor and all of its descendants.

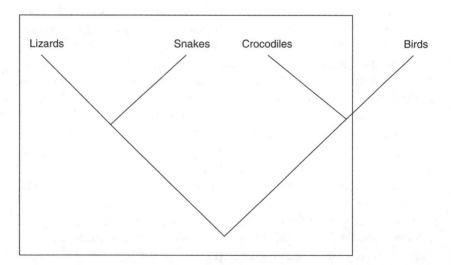

Figure 2.3   The taxon Reptilia contains lizards, snakes, and crocodiles, but does not contain birds. Reptilia is a paraphyletic taxon: it contains an ancestor and some but not all of its descendants.

B splits into D and E, resulting in two new monophyletic groups, D and E. The process of anagenesis, on the other hand, gives rise to lineages that are paraphyletic. According to evolutionary taxonomists, birds have significantly diverged from lizards, snakes, and crocodiles, so birds are excluded from the taxon Reptilia. If birds are excluded from Reptilia, then Reptilia does not contain all the descendants of its most recent ancestor and is therefore paraphyletic.

The difference between monophyletic and paraphyletic taxa is an important one that divides evolutionary taxonomists from cladists. While evolutionary taxonomists construct both monophyletic and paraphyletic taxa, cladists argue that only monophyletic taxa should be posited. Cladists believe that classification should be strictly genealogical. However, membership in a paraphyletic taxon is not defined merely by common ancestry but also by how much divergence has occurred among an ancestor's descendants. Because both of these factors are used for constructing paraphyletic taxa, cladists reject the existence of paraphyletic taxa. Membership in Reptilia, for example, requires being descended from a common ancestor. Yet according to evolutionary taxonomists, that common ancestry is not sufficient for membership in Reptilia: birds have significantly diverged from reptiles, so they should be excluded from Reptilia. Cladists, on the other hand, deny the existence of the paraphyletic taxon Reptilia.

Why do evolutionary taxonomists desire classifications that capture both monophyletic and paraphyletic groups? Evolutionary taxonomists maintain that each type of group is the result of a particular sort of evolutionary process, namely, cladogenesis and anagenesis. They argue that to ignore one of these processes in classification is to neglect a significant part of evolution. Hence, the need exists for classifications containing both monophyletic and paraphyletic taxa. Evolutionary taxonomists bolster their contention by citing Darwin. In the *Origin*, Darwin makes an observation that seems very much aligned with the thinking of evolutionary taxonomists:

> I believe that the *arrangement* of the groups within each class, in due subordination and relation to the other groups, must be strictly genealogical in order to be natural, but that the *amount* of difference in the several branches or groups, though allied in the same degree in blood to their common progenitor, may differ greatly, being due to the different degrees of modification which they have undergone, and this is expressed by the forms being ranked under different genera, families, sections, or orders (1859, 420).

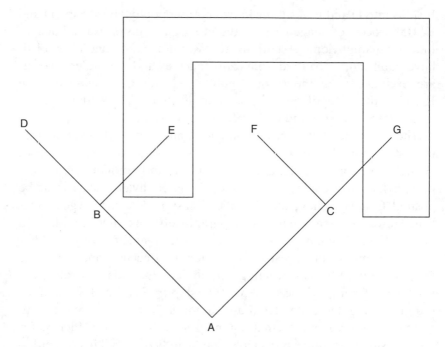

Figure 2.4   The taxon containing only E and G is polyphyletic. E and G share similar traits (homoplasies) that were not present in their common ancestor, A, but evolved independently in E and G.

We have seen the sorts of taxa that are provided in the classifications of evolutionary taxonomists. It is equally important to see the types of taxa they believe should be excluded, namely, polyphyletic taxa. Such taxa consist of lineages whose organisms share characters that were not present in their common ancestor but later evolved by independent evolution. Figure 2.4 provides an example of a polyphyletic taxon. The polyphyletic taxon highlighted contains only lineages E and G. Furthermore, E and G contain similar characteristics not found in their common ancestor, A, but which independently evolved in E and G. Another way of distinguishing polyphyletic taxa from monophyletic and paraphyletic taxa is the different sorts of characters used to identify those taxa. Polyphyletic taxa contain organisms that have *homoplasies*: traits shared by different lineages but not present in their common ancestor. Monophyletic and paraphyletic taxa contain organisms that have *homologies*: traits shared by different lineages that were passed down from a common ancestor.

Homologies, themselves, come in two states, and that division is used to distinguish monophyletic from paraphyletic taxa. *Ancestral* (or *plesiomorphic*) character states are present in the ancestor of an entire group. *Derived* (or *apomorphic*) character states arise later in a group. In Figures 2.5a and 2.5b, X is the ancestral state of a character for the groups illustrated, and X' is the derived state of that character. Derived character states but not ancestral states serve as evidence of monophyly. In Figure 2.5a, the derived character, X', occurs in the common and unique ancestor of taxa D and E. So the occurrence of X' in D and E indicates that D, E, and their most recent common ancestor form a monophyletic group. In Figure 2.5b, the ancestral state, X, occurs in E, F, G, but not D. Because X does not occur in all the descendants of that group's ancestor, E, F, G, and their ancestor form a paraphyletic group.

We will return to the distinction between ancestral and derived homologous character states in Section 2.3 when we discuss cladism. For now, let us focus on the difference between homoplasies and homologies. That difference is important for evolutionary taxonomists because they cite it in justifying their belief that monophyletic and paraphyletic taxa but not polyphyletic taxa should be classified. Monophyletic and paraphyletic taxa consist of traits that were passed down from a common ancestor. Taxa that contain homologous traits are parts of a single genealogical entity whereas taxa containing homoplastic similarity are the result of independent events and not a common evolutionary source. From a genealogical perspective, polyphyletic taxa are aggregates of independently working processes. For the evolutionary taxonomists, polyphyletic taxa are not natural taxa and should be excluded from classification. A major task of evolutionary taxonomists has been to remove polyphyletic taxa from earlier classifications.

One last item concerning evolutionary taxonomy should be addressed. When is divergence within a lineage significant enough to recognize a new taxon? This is a pressing question because whether an instance of divergence counts as significant affects the classification of taxa. Consider the case of birds and reptiles (Figure 2.3). The taxon Reptilia includes lizards, snakes, and crocodiles but not birds so long as birds significantly diverged from lizards, snakes, and crocodiles. Originally, evolutionary taxonomists suggested that significant divergence depended on less similarity in the members of two taxa than the similarity found within each taxon (Ridley 1986, 29). However, evolution-

Taxa   D   E   F   G

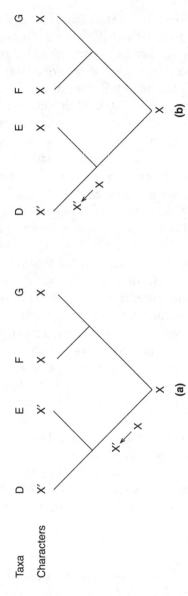

Figure 2.5   X is the ancestral state of the character in question. X' is the derived state of the character. Derived character states but not ancestral character states serve as evidence of monophyly. (a) The derived character state, X', occurs in the common and unique ancestor of taxa D and E and indicates that D, E, and their most recent common ancestor form a monophyletic group. (b) The ancestral state, X, occurs in E, F, G, but not D. Because X does not occur in all the descendants of that group's ancestor, E, F, G, and their most recent common ancestor form a paraphyletic group.

ary taxonomists have rejected defining significant divergence in terms of overall similarity. For one, overall similarity breaks up groups of genealogically connected organisms, not to mention the different life stages of a single organism. Examples of the first are classifications containing only the males of different species (Mayr 1963, 32); examples of the second are classifications containing only the juveniles of different species (Wiley 1981, 33). Evolutionary taxonomists also rally against the criterion of overall similarity because they see it as a throwback to essentialism. They charge that basing classifications on overall similarity assumes that taxa have underlying essences (Ridley 1993, 374). As we shall see in the next chapter, one of the founders of evolutionary taxonomy, Mayr, is very much against an essentialistic approach to biological classification.

More recently, evolutionary taxonomists have offered an account of significant evolutionary divergence that does not rely on overall similarity. They suggest that significant divergence occurs when a lineage enters a new environment and acquires a radically new set of adaptations. For instance, Mayr (1981[1994], 286) writes, "Whenever a clade (phyletic lineage) enters a new adaptive zone that leads to a drastic reorganization of the clade, greater taxonomic weight may have to be assigned to the resulting transformation than to proximity of joint ancestry." An adaptive zone, according to Mayr (1988, 135), is "any kind of environment to which an organism can become adapted." For evolutionary taxonomists, the taxon Aves is an example of a portion of a lineage entering a new adaptive zone and acquiring a new suite of adaptations (Figure 2.3). The ancestors of birds entered a new adaptive zone relative to that occupied by land or water-based reptiles. That transition led to the acquisition of a radically new set of adaptations (wings being one of them).

Evolutionary taxonomists have a name for those portions of lineages that enter a new adaptive zone and are drastically reorganized. They call such lineages "grades." According to Mayr (1976, 450):

> All members of a grade are characterized by a well-integrated adaptive complex. . . . In the history of the vertebrates we know many such cases of the formulation of successful new grades, such as the sharks, the bony fishes, the amphibians, reptiles, birds, and mammals. Each of these is characterized by a certain type of adaptation to the environment.

Despite initial appearances, grades are not a throwback to essentialism or classification by overall similarity. Mayr is not asserting that all and only the members of a grade contain a certain adaptation. Moreover, he is not maintaining that the organisms in a grade are more similar to one another than they are to the members of a different taxon. Yet as many critics of evolutionary taxonomy argue, what is positively asserted by saying that the organisms of a taxon form a grade needs further clarification (see, for example, Eldredge and Cracraft 1980).

2.2   PHENETICISM

In large part, pheneticism is a reaction to the practices of evolutionary taxonomy and earlier schools of classification. A major defect pheneticists see in evolutionary taxonomy is that school's use of the homology/homoplasy distinction. Consider the following argument provided by Sokal and Sneath (1963, 7–8) in the first text on pheneticism. Evolutionary taxonomists use only homologies in constructing classifications. Yet to know whether a character is a homology rather than a homoplasy requires knowing the genealogy of the taxa in question. Rarely do biologists have access to such information. So in most cases, evolutionary taxonomists must assume which characters are homologies and which are homoplasies. Sokal and Sneath (1963, 7) conclude that "the existence of [taxon] A as a natural (or 'monophyletic') group defined by character complex X has been *assumed* but *not demonstrated*." Pheneticists believe that the distinction between homology and homoplasy exists. However, they maintain that it is not practically feasible for taxonomists to distinguish homologous and homoplastic similarity in their work.

Pheneticists also argue that the homology/homoplasy distinction should be rejected because no type of character should be preferred when constructing classifications; all characters should carry equal weight. Their motivation for treating all traits equally is twofold. First, pheneticists point out that the selection of characters used in classification is often guided by theory. However, biological theory changes over time. Many of the theoretical assumptions accepted by Aristotle and Linneaus were rejected by Darwin. Some contemporary biologists (for example, Gould and Eldredge) reject parts of Darwinism. Who knows what parts of contemporary evolutionary theory will be cast aside in the future? If classification is tied to theory, then incorrect theory will lead to incorrect classification. Pheneticists suggest that this has occurred in the past and should be avoided in the future (Cain

1958, 1959a). Accordingly, they argue that theoretical considerations should not dictate which types of characters should be used in classification. (We will evaluate the prospects of constructing theory-neutral classifications in Chapters 3 and 5.)

The other reason pheneticists provide for taking all characters on a par stems from what they see as the general purpose of biological classification. According to Sokal, taxonomists should attempt to provide classifications that serve as a basis for making various types of hypotheses (1985[1994], 244). Such hypotheses should not be limited to those used by evolutionists but should include hypotheses used by various types of biologists, such as embryologists, anatomists, and ecologists. Given these considerations, pheneticists argue that classifications should be based on all types of characters rather than merely those that highlight evolutionary relations. As Sneath and Sokal (1973, 10) write, "[f]rom the point of view of biology in general, it is probably of more interest to describe the overall similarity of organisms than their branching sequences."

The desire for theory neutral all-purpose classifications causes pheneticists to adopt a strict empiricist philosophy. Taxonomists should simply observe and record the resemblances found in nature; no theoretical assumptions should bias the selection of characters observed. In other words, taxonomists should approach the study of nature as blank slates on which nature can impress itself. According to Colless (1967), a phenetic classification refers "only to the observed properties of such entities, without any reference to inferences that may be drawn *a posteriori* from the patterns displayed. Such classification can, and to be strictly phenetic *must*, provide nothing more than a summary of observed facts."

So how do pheneticists suggest carrying out this strict empiricist approach to classification? The methodology of pheneticism is displayed in a flow chart on the first page of Sokal and Sneath's (1963) text (Figure 2.6). First, organisms are gathered (or observed) and their characters are recorded. No character is preferred over another, and as many characters are recorded as possible. Classifications based on only several characteristics may be biased by the particular nature of those characteristics. For example, a classification based on immunoglobulin concentration would probably differ from one based on body length (Ridley 1986, 38–9). To avoid the idiosyncrasies of particular characters, pheneticists recommend performing phenetic studies involving dozens if not hundreds of characteristics.

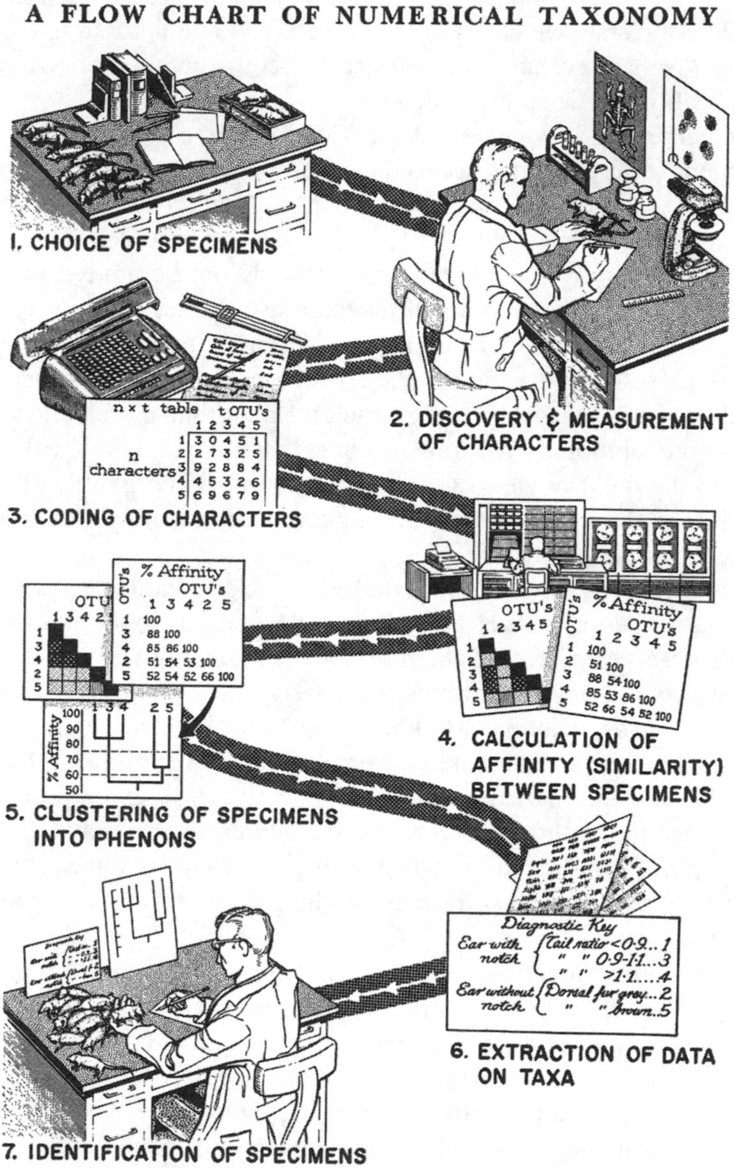

A FLOW CHART OF NUMERICAL TAXONOMY

Figure 2.6   A flow chart highlighting the methodology of numerical taxonomy, *circa* 1963. Reprinted with permission from R. Sokal and P. Sneath, *The Principles of Numerical Taxonomy* (San Francisco: W. H. Freeman, 1963, p. xviii).

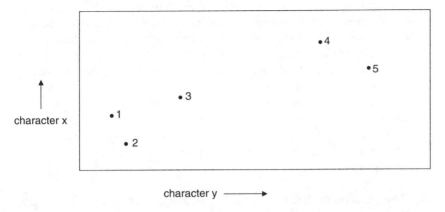

Figure 2.7   A two-dimensional phenetic space, one dimension for character X and another dimension for character Y. Five operational taxon units (OTUs), 1 through 5, are plotted according to the occurrence of characters X and Y in those units.

The result of these observations is a list of characters found among the various "operational taxonomic units" (OTUs) under study. OTUs are the provisional taxonomic units being classified; they might be species or they might be higher taxa. What pheneticists take to be base OTUs in classifications, namely species, is an issue we will consider in Section 2.5. The concern here is how pheneticists construct classifications given data about OTUs. The step between data and classification is "the calculation of affinity (similarity) between specimens." Some characters may occur in all the taxa under consideration, so those characters are ignored. The remaining characters are used to determine how similar OTUs are to one another. This is done with statistical techniques. A multidimensional space, one dimension for each characteristic, is envisioned, and OTUs are placed in that space according to their having those characters. Consider a simple case involving two characters and hence two dimensions (Figure 2.7). Five taxa are plotted according to the occurrence of two characters within those taxa. Pheneticists then measure the "distance" between taxa in phenetic space. That distance is a measure of the similarity among those taxa.

The next step is to measure the phenetic distances among OTUs and to construct an appropriate classification. Sneath and Sokal (1973, 7) suggest that this procedure is akin to reading geographic features off a topographic map:

numerical methods . . . are methods for establishing and defining clusters of mutually similar entities from the $t \times t$ resemblance matrix. These clusters may be likened to hills and peaks on a topographic chart, and the criteria for establishing the clusters are analogous to the contour lines of such a map. Rigid criteria correspond to high elevation lines that surround isolated high peaks – for example, species groups in a matrix of resemblances between species. As the criteria become relaxed the clusters grow and become interrelated in the same way that isolated peaks acquire broader bases and become connected to form mountain complexes and eventually chains, with progress from higher- to lower-contour lines.

Applying this metaphor to Figure 2.7, we can see a classification emerging. Taxa 1 and 2 are clustered into one peak, as are taxa 4 and 5, while taxon 3 stands alone. Assuming that our OTUs are species, perhaps each of these "peaks" represents a distinct genus: taxa 1 and 2 belong to one genus, taxon 3 its own, and taxa 4 and 5 another genus. We also see that taxon 3 is closer to taxa 1 and 2 than taxa 4 and 5. So perhaps taxa 1, 2, and 3 form a family distinct from the family containing taxa 4 and 5. From such considerations, a hierarchical classification is constructed. What distances between OTUs and clusters of OTUs count as differences among species versus differences among higher taxa is, according to pheneticists, an arbitrary decision (Sneath and Sokal 1973, 7).

Pheneticists use all types of characters in constructing classifications. So, unlike evolutionary taxonomists, they use both homoplasy and homology as taxonomic data. This difference between phenetics and evolutionary taxonomy gives rise to another. While evolutionary taxonomists rule out the inclusion of polyphyletic taxa in classifications, the phenetic acceptance of homoplastic similarity allows such taxa. Consider the following example. Taxa A, B, and C have a common ancestor, X (Figure 2.8). Two types of characters are found in those taxa: homologies (h_1, h_2, and h_3), and homoplasies (a_1, a_2, a_3, a_4, and a_5). Suppose that characters come in two states: either a character occurs (signified by "1"), or it does not (signified by "0"). The distribution of characters found in A, B, and C is the following:

	A	B	C
h_1	1	1	1
h_2	0	1	1
h_3	0	1	1

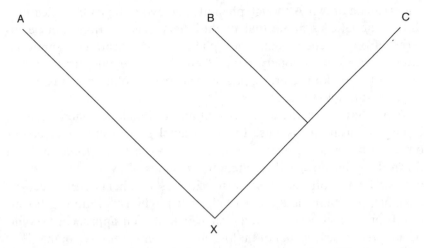

Figure 2.8   The phylogenetic relations among four taxa, where X is the common ancestor of A, B, and C.

a_1	1	0	1
a_2	1	0	1
a_3	1	0	1
a_4	1	0	1
a_5	1	0	1

Character h_1 occurs in all three taxa, so it cannot be used by pheneticists to classify taxa. Taxa B and C share two homologous characters (h_2 and h_3) not found in taxon A. Taxa A and C have 5 characters (a_1–a_5) that rose independently after their common ancestor X. These characters are homoplasies. On the basis of overall similarity, A and C are more similar to one another than either is to B. Consequently, pheneticists should posit the existence of a taxon that includes A and C, but excludes B and their common ancestor X. Such a taxon is polyphyletic. An evolutionary taxonomist, on the other hand, would argue that characters a_1 through a_5 do not serve as evidence for taxonomic relations, whereas h_2, and h_3 do. Consequently, A and C do not form a higher taxon, but B and C do.

Despite such examples, pheneticists believe that with the appropriate assumptions, phenetic classifications can be used to construct evolutionary ones (Sneath and Sokal 1973, 5). Recall that pheneticists see their classifications as general purpose ones from which various inferences can be drawn. Although phenetic and evolutionary classifications differ, pheneticists argue that evolutionary classifications can be inferred from phenetic data, but phenetic classifications cannot be inferred from evolutionary data. Pheneticists see this distinction as another reason for preferring pheneticism over evolutionary taxonomy (Sneath and Sokal 1973, 10).

As we have seen, the development and rationale for pheneticism is in large measure a response to the principles of evolutionary taxonomy. Evolutionary taxonomists have not taken such criticisms lying down. They have replied to phenetic critiques of evolutionary taxonomy and have offered their own critiques of pheneticism (see, for example, Mayr 1965 and Mayr 1969, 203ff.). While evolutionary taxonomists and pheneticists argued over which of their approaches should be adopted, a new school of biological taxonomy was developing. That school was cladism.

2.3 PROCESS CLADISM

Willi Hennig, the founder of cladism, first introduced this school of taxonomy in his 1950 treatise, *Grundzuge einer Theorie der Phylogenetichen Systematik*. However, most English-speaking taxonomists did not learn about cladism until the 1960s when Hennig starting publishing his work in English.

Hennig's general aim was to find a system of classification that would serve as "the general reference system" for all biological disciplines (1966, 7, 9). Hennig (1965, 1966) considers two contending systems of classification: one based on phylogeny and the other stemming from ideal morphology. Ideal morphology was popular among German biologists in the first half of the twentieth century. According to ideal morphology, nature consists of basic patterns called "bauplans," and the goal of ideal morphology is to discover those bauplans by merely observing morphological traits. Ideal morphologists see no need to know anything about the evolutionary history of the organisms under study. Hennig's initial argument for preferring phylogenetic classifications is in large part a negative one against ideal morphology. Hennig writes that ideal morphologists mistakenly believe that we can construct scientific classifications without making any theoretical assumptions about evolutionary relations (1966, 8, 11–12). Furthermore, he argues that ideal morphologists wrongly assume that the primary relations among organisms are similarity relations. Alternatively, Hennig suggests that the primary relations among organisms are causal ones such as genealogical and embryological relations (1966, 12).

It is not a long stretch to see that some of Hennig's criticisms of ideal morphology apply to pheneticism. Pheneticists also believe that biological classifications should be made independent of any theoretical assumptions concerning evolutionary relations. Many followers of Hennig have criticized pheneticism for making such assumptions. We will turn to some of those criticisms in Chapters 3 and 5. Cladists are also unhappy with the reliance that evolutionary taxonomists place on a certain type of similarity – that the organisms of an evolutionary grade share a similar adaptive complex. Of course, the principles of evolutionary taxonomy are quite different from those of pheneticism. Evolutionary taxonomists rely heavily on theory in constructing classifications, and they certainly do not think that similarity is the only criterion for making taxonomic decisions – information about genealogy is essential (see Section 2.1). We will touch on the debate between evolutionary taxonomists and cladists in this and later chapters. But first let us look at the positive tenets of cladism.

Hennig wanted to develop a system of classification that unambiguously captures phylogeny. Hennig's system highlights the branching relations on the tree of life. That is reflected in the name of the taxonomic school Hennig founded. "Cladism" is based on the Greek word for branch. Two taxa should be grouped together when they originate in the same branching event. To put it slightly differently, two taxa originate in the same branching event if they share a more recent common ancestor than either shares with a third taxon. So a major goal of cladism is to construct classifications that show "recency of common ancestry" (Hennig 1965[1994], 258). Consider three taxa, A, B, and C. A and B should be grouped together if they share a more recent common ancestor than either shares with C. If they do, A and B are "sister taxa."

The cladistic desire that classification reflect recency of common ancestry can be described in another way. In the previous two sections, we saw a distinction among three types of taxa: monophyletic, paraphyletic, and polyphyletic taxa. Recall that a monophyletic taxon contains an ancestor and all and only its descendants. This notion of monophyletic taxa is defined just by recency of common ancestry. In Figure 2.2 the taxon consisting of H, I, and G is monophyletic because H and I share a more recent common ancestor, G, than either shares with any other taxon. Similarly, H, I, G, F, and C form a monophyletic taxon because H, I, G, and F together share a more recent common ancestor, C, than any of them shares with another taxon. Monophyletic taxa reflect common ancestry and nothing else. They are exactly the sorts of taxa cladists want to capture by classifications. As a result, cladists require that classifications contain only monophyletic taxa.

It is not hard to see why cladists do not recognize either paraphyletic or polyphyletic taxa. Recall that evolutionary taxonomists place lizards, snakes, and crocodiles in the paraphyletic taxon Reptilia (Figure 2.3). Crocodiles and birds share a more common recent ancestor than crocodiles share with any of the other taxa in Reptilia. Still, evolutionary taxonomists distinguish Reptilia from birds because they believe that the organisms in those two groups have significantly different adaptive zones. The cladist does not recognize the paraphyletic taxon Reptilia because the division between Reptilia and birds ignores the recent common ancestor that crocodiles and birds share. Similarly, cladists do not recognize polyphyletic taxa. Polyphyletic taxa consist of lineages whose organisms share characters that were not present in their common ancestor but only later evolved independently. In Figure 2.4, E and G form a polyphyletic taxon because of their phenotypic similarities. Cladists reject polyphyletic taxa because they do not reflect recency of common ancestry. Lineages E and G form a polyphyletic taxon, but each shares a more recent common ancestor with a lineage outside that taxon. G and F, for instance, share a more recent common ancestor, C, than G shares with E. Accepting a polyphyletic taxon such as the one containing lineages E and G conflicts with the cladistic prescription to classify by common ancestry.

Having seen that cladists aim to classify taxa by recency of common ancestry, a question arises: How do cladists determine recency of common ancestry? They do so by using information concerning the similarities found among the organisms in the taxa being studied. However, cladists argue that not any type of similarity provides evidence for recency of common ancestry; only a particular type of character will do the trick. Recall the two types of characters discussed in the previous sections: homoplasies and homologies. A character is a homoplasy when it occurs in different lineages but not the common ancestor of those lineages. Thus homoplastic similarity does not reflect common ancestry.

Having ruled out homoplasies as evidence of common ancestry, cladists are left with constructing classifications using homologies. Cladists contend that only certain types of homology can be used. Derived similarities (or *synapomorphies*) are unique to an ancestor and all its descendent taxa. Ancestral similarities (or *symplesiomorphies*) are characters found in an ancestor and some but not all of its descendants. Cladists maintain that only synapomorphies can serve as evidence of recency of common ancestry. Their point can be illustrated with the example of Reptilia (Figure 2.3). Winglessness is plesiomorphic in lizards, snakes, crocodiles, and the common ancestor of birds and Reptilia. Winglessness is not found in birds. If we classify taxa by this symplesiomorphy, as evolutionary taxonomists do, we end up with the taxon Reptilia. Yet the paraphyletic taxon Reptilia does not reflect recency of common ancestry: crocodiles share a more recent common ancestor with birds than they do with either lizards or snakes. So according to cladists, symplesiomorphies cannot be used as evidence for recency of common ancestry. Synapomorphies, on the other hand, are clear evidence of common ancestry. Consider the example in Figure 2.5a. X′ is apomorphic in D and E but not in F or G. If we have evidence that x′ is apomorphic in only D and E, we have evidence that D and E share a more recent common ancestor than either shares with any other taxon.

We have seen that cladists use synapomorphies rather than homoplasies or symplesiomorphies for constructing classifications. But how do cladists determine which similarities are synapomorphies? This question can be broken into two parts. How do cladists distinguish homologies from homoplasies? How do they distinguish synapomorphies from symplesiomorphies? Cladists use various criteria for distinguishing homologies from homoplasies. Here are two (from Ridley 1993, 454–5). One is to require that instances of a homologous character have the same fundamental structure and not merely a superficial similarity. The wings of birds and bats, for example, violate this criterion. Though their wings look similar on the surface, they are supported by different digits and are composed of different materials. A second criterion requires that instances of a homologous character be similar not only in its adult form but also at various stages of embryonic development. Barnacles and limpets have similar adult characteristics (hard external armor, feeding through a hole in the shell, etc.), but look quite different as larvae. Consequently, the similar adult characters of barnacles and limpets are considered homoplastic.

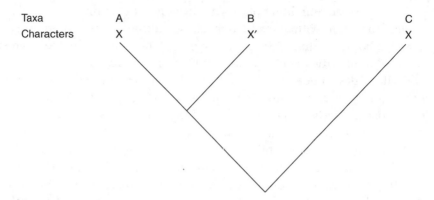

Figure 2.9 An example of outgroup comparison. X and X′ are homologous states of the same character. X occurs in A and X′ occurs B. For taxa A and B, which character state is ancestral and which is derived? If C is the outgroup and X occurs in C, then X is the ancestral character state and X′ is the derived state.

Having distinguished homoplasies from homologies, the cladist turns to distinguishing which homologies are synapomorphies and which are symplesiomorphies. The cladistic techniques for doing so are varied, but the most common technique is outgroup comparison. Suppose a biologist has two species, A and B. X and x′ are different states of the same character, and x occurs in A while x′ occurs in B (Figure 2.9). Both x and x′ are considered homologies, but which is the derived state and which is the ancestral state? Outgroup comparison works by observing a taxon closely related to A and B, and determining the state of the character in that taxon. The character state found in that third taxon – the outgroup – should be considered ancestral. In our example, C is the outgroup and character state x occurs in that species. Consequently, x should be considered ancestral while x′ should be considered derived.

What is the justification for outgroup comparison? It is the principle of parsimony (Ridley 1986, 61; Sober 1993, 177). If the character state of the outgroup is ancestral, rather than derived, then fewer evolutionary changes are required to arrive at the observed character distribution. In Figure 2.9, if x is considered the ancestral state, then at least one evolutionary change is required for taxon B to have x′. On the other hand, if x is considered the derived state, then at least two evolutionary changes must occur: one in the branch leading to C, and one in the branch leading to taxon A. Thus, the assumption that the outgroup's character is the ancestral state is more parsimonious than assuming that it is derived.[3]

Once cladists have determined which character states are ancestral and which are derived, the next stage of analysis is to construct a *cladogram* using observed synapomorphies. A cladogram is a branching diagram representing hierarchical relations among a set of taxa based on observed synapomorphies. The resultant taxa are called "clades." Using an example from Hickman, Roberts, and Larson (1997, 200), suppose we want to construct a cladogram for four vertebrates – bass, lizard, horse, and monkey (Figure 2.10). Four characters are used to construct the cladogram: having (or lacking) vertebrae, four legs, hair, and mammary glands. The organisms in the outgroup, Amphioxus, lack all of these characters, so their absence is the ancestral state of these characters (plesiomorphic) for vertebrates. Having vertebrae is apomorphic for bass, lizard, horse, and monkey; thus those taxa form a single taxon. Having four legs is apomorphic for lizard, horse, and monkey; hence those three taxa form a clade relative to bass. Having hair and mammary glands is apomorphic for horse and monkey; hence those taxa form a less inclusive clade.

It is important to note that cladograms are not *phylogenetic trees*. Cladograms merely indicate the nesting of clades using observed synapomorphies as evidence. Phylogenetic trees, on the other hand, represent branches on real lineages in the evolutionary past. To obtain a phylogenetic tree from a cladogram we must add information concerning ancestor/descendant relationships and rates of evolutionary change (Eldredge and Cracraft 1980, 212–15; Sober 1988, 22–4). Nevertheless, cladograms are often used as first approximations of phylogenetic trees.

In the above example of vertebrates, constructing a cladogram seemed simple. However, the move from observations of synapomorphies to cladograms can be problematic. Suppose a biologist wants to classify three taxa, A, B, and C, and she has three characters, x, y, and z. (This is a simplified example; in the vast majority of cases, cladists classify using dozens of characters.) Furthermore, suppose that each character comes in one of two states, 0 and 1: 0 is ancestral and 1 is derived. The distribution of characters in the three taxa is the following:

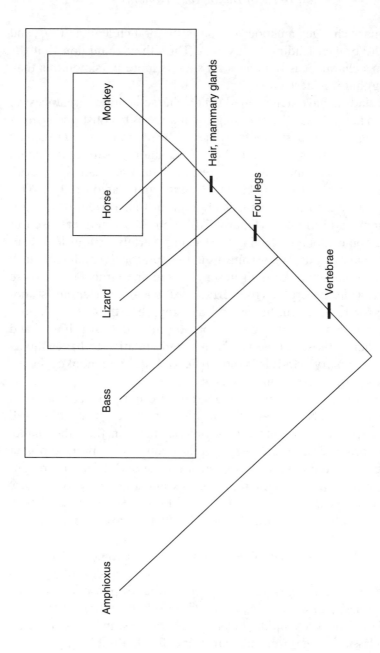

Figure 2.10 A cladogram of four vertebrates: bass, lizard, horse, and monkey. Amphioxus is the outgroup. Adapted from Hickman, Roberts, and Larson 1997, 200.

	A	B	C
x	1	1	0
y	1	1	0
z	0	1	1

This distribution gives rise to the two conflicting cladograms (Figures 2.11a and 2.11b). According to one cladogram, A and B are more closely related to each other than either is to C (Figure 2.11a). In the other, B and C are more closely related (Figure 2.11b). What is a cladist to do in such a situation?

First off, the cladist is very clear concerning what has gone wrong here: some of the suggested synapomorphies are not really homologies but homoplasies. If the first cladogram is correct (Figure 2.11a), then z is a character that has evolved at least twice: once in the branch leading to C and once in the branch leading to B. On the other hand, if the second cladogram (Figure 2.11b) is correct, then x and y are homoplastic. If that is the case, then x and y have each evolved at least twice: both on the branch leading to A and both on the branch leading to B. The task for the cladist is to infer which of the putative homologies are homoplasies. To do this, the cladist again employs the principle of parsimony. Here she chooses the phylogeny that requires the minimal number of evolutionary changes to arrive at the given character distribution. In the example under consideration, the principle of parsimony counsels choosing the phylogeny represented by the first cladogram. The phylogeny captured by the first cladogram minimally requires four evolutionary changes (they are represented by slash-marks on the cladogram) whereas the phylogeny captured by the second cladogram minimally requires five evolutionary changes.

Cladists offer various justifications for their use of parsimony in discerning synapomorphies. Some cladists argue that evolution itself is parsimonious. The claim here is not that evolution takes the minimal number of steps to arrive at observed character distributions. Instead, a more modest claim is put forth: more often than not evolution gives rise to similarities that require at least one change rather than similarities that require at least two changes. For example, Ridley (1986, 82) writes:

> We have no strong principle telling us to use the method of maximum parsimony; we only have a weak one. It is that evolution is relatively improbable. It is unlikely enough that all the mutations should arise and be selected for one character in one species, but that similar events

Taxa   A   B   C   A   B   C

Characters

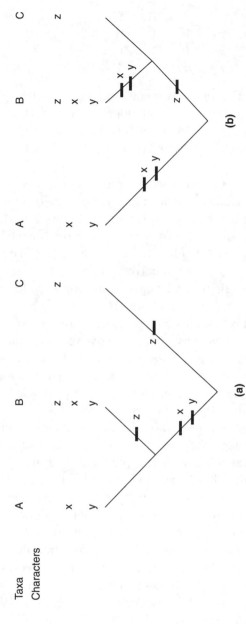

(a)   (b)

Figure 2.11 Two different cladograms of the same taxa. (a) If taxa A and B are more closely related to each other than either is to C, then the occurrence of X and Y in taxa A and Y in taxa A and B is homologous and the occurrence of Z in taxa B and C is homoplastic. (b) If taxa B and C are more closely related to each other than either is to A, then the occurrence of Z in taxa B and C is homologous and the occurrence of X and Y in taxa A and B is homoplastic.

should take place independently in another is even more improbable. Shared characters are therefore more likely to be due to common ancestry than to convergence.

Other cladists justify their use of parsimony on more philosophical grounds. They argue that we should prefer phylogenetic hypotheses that posit the fewest number of homoplasies because to assume that a putative synapomorphy is homoplastic is to make an *ad hoc* assumption (that is, an assumption lacking empirical evidence). According to cladists that offer this line of reasoning (for example, Eldredge and Cracraft 1980, 70, and Wiley 1981, 111), biologists should minimize the number of *ad hoc* assumptions they make regardless of whether evolution is parsimonious. Although cladists have written extensively on the proper justification for the use of parsimony, the issue is far from settled. (See Sober 1988 for a summary and critical analysis of the literature on parsimony.)

Cladists offer other methods besides parsimony for handling conflicting data, but we need not explore those methods here. Enough has been said to provide a sufficient introduction to process cladism for the purposes of this book. Let us now turn to the second and more recent branch of cladism: pattern cladism.

2.4 PATTERN CLADISM

The cladists cited above, Hennig, Ridley, and Wiley, see an intimate connection between evolutionary theory and cladism. Cladistic classifications should highlight a particular aspect of evolution, namely, the grouping of taxa by recency of common ancestry. Synapomorphy serves as evidence that a group of taxa shares a common ancestor. Not all cladists, however, believe that cladistic classifications highlight common ancestry. A newer breed of cladists believe that cladism should be decoupled from any theoretical assumptions about evolution, even the assumption that taxa are genealogical entities. Consider the following quotations from a leading advocate of this branch of cladism:

> As the theory of cladistics has developed, it has been realized that more and more of the evolutionary framework is inessential, and may be dropped. . . . Platnick refers to the new theory as "transformed cladistics" and the transformation is away from a dependence on evolutionary theory (Patterson 1980 [1982, 118]).

As I understand it, cladistics is theoretically neutral so far as evolution is concerned. It has nothing to say about evolution. You don't need to know about evolution, or believe in it, to do cladistic analysis. All that cladistics demands is that groups have characters, and that the groups are nonoverlapping (Patterson 1981).

This school of cladism has been given a couple of names. One is "transformed cladism," as seen in the first quotation. Another is "pattern cladism." I will use the latter because it nicely highlights the difference between this newer form of cladism and the more traditional, evolutionary based approach to cladism. Following others, I have called the more traditional branch of cladism "process cladism." Process cladists believe that synapomorphies are the result of common descent; thus classifications based on such similarity reflect genealogical relations among taxa. Pattern cladists, on the other hand, believe that taxonomists only have access to patterns of characteristics, and consequently, taxonomists should not infer that those patterns are the result of any particular process. Consider the following quotation from the leading text on pattern cladism:

To state that a cladogram is a synapomorphy scheme invites the rejoinder that a cladogram must, therefore, be a phyletic concept. Not so, for by "synapomorphy" we mean "defining character" of an inclusive taxon. True, all defining characters, in the phyletic context, may be assumed to be evolutionary novelties. But making that assumption does not render it automatically true; nor does it change the characters, the observations on which characters are based, or the structure of the branching diagram that expresses the general sense of the characters: i.e., that there exist certain inclusive taxa ... that have defining characters (Nelson and Platnick 1981, 324).

Why do pattern cladists maintain that cladistic classification should be devoid of any evolutionary assumptions about the taxa being classified or the characters being used as evidence? It is in large part because they want to promote a cladistic approach to classification that is theory neutral and does not depend on any evolutionary assumptions. Pattern cladists provide a variety of reasons for preferring a theory-neutral approach to classification. One is the reason that pheneticists provide in advocating a theory-free taxonomic approach: the preference for empiricism. Biological theories come and go, but the observations of organisms used for classification remain constant.

Pattern cladists do not want to infect those observations with "futile theorizing" (Nelson and Platnick 1981, 151).

A second reason pattern cladists prefer theory-neutral classifications follows another line of phenetic thinking: theory-free classifications are necessary if classifications are to serve as evidence for biological theory. Platnick (1979, 539), for instance, writes that "if classifications (that is, our knowledge of patterns) are ever to provide an adequate test of theories of evolutionary processes, their construction must be independent of any particular theory of process." A third and perhaps more fundamental reason pattern cladists prefer theory-free classifications is the belief that theories about process can be formulated only *after* classifications concerning patterns are obtained. According to Nelson and Platnick (1981, 171), "[p]attern analysis is, in its own right, both primary and independent of theories of process, and is a necessary prerequisite to any analysis of process."

Pattern cladism and pheneticism share a common set of motivating principles. But surely there is a divide between the two schools given that pattern cladism falls within the more general school of cladism. Platnick (1979, 544) suggests that the main difference between pattern cladism and pheneticism is that pheneticists accept all similarities as evidence whereas pattern cladists do not. Recall that pheneticists classify according to overall similarity. Often some similarities will contradict the classification given by the most overall similarity. Suppose, for example, taxa A and B share three similar characters while taxa B and C share four similar characters. Those two sets of similarities give rise to different classifications. The pheneticist chooses the classification that reflects the highest congruence of similarity while allowing that some similarities may go against that classification. According to Platnick (1979, 544), the pattern cladist insists that all observed characteristics give rise to one classification. Any "noncongruent" characters – those that give rise to conflicting classifications – are dismissed as not really being "characters." So for pheneticists, all similarities serve as evidence, whereas for pattern cladists, only some similarities are evidence. But which similarities are those?

For process cladists, only synapomorphies can be used to construct classifications. A synapomorphy, according to process cladists, is a derived similarity that originates in an ancestral taxon and is passed down to its descendant taxa. The pattern cladist cannot adopt the process cladist line here, given the pattern cladist's view that classifications should be theory neutral. Accordingly, pattern cladists provide their own definition of "synapomorphy." They describe synapomorphy as follows:

> But *all* of Hennig's groups correspond by *definition* to patterns of synapomorphy. Indeed, Hennig's trees are frequently called synapomorphy schemes. The concept of "patterns within patterns" seems, therefore, an empirical generalization largely independent of evolutionary theory, but, of course, compatible with, and interpretable with reference to, evolutionary theory. The concept rests on the same empirical basis as all other taxonomic systems (the observed similarities and differences of organisms) (Nelson and Platnick 1981, 141–2).

For the pattern cladist, only nested patterns of synapomorphy can be observed in nature. Moreover, we can observe them with no theoretical assumptions concerning the processes behind them. Cladograms, and their corresponding classifications, display nothing more than the patterns of synapomorphies observed in nature. This is in stark contrast to the process cladist assumption that observed patterns of synapomorphies serve as evidence for, and can only be understood as the result of, certain evolutionary processes. Once again, the desire for theory-neutral classifications is pitted against the desire for theory-dependent classifications.

This concludes our survey of the four primary schools of biological taxonomy. Several points of contention nicely contrast these schools and their systematic philosophies. The first is the type of similarity each allows for constructing classifications. For pheneticists, any similarity among organisms serves as evidence for constructing classifications, provided those similarities are not universal to all organisms. Evolutionary taxonomists allow only similarities that are homologies – similarities that are the result of common ancestry and not independent evolution. Cladists place the further restriction that only synapomorphies count. Synapomorphies are evidence of complete branches or clades.

Disagreement over which similarities count as evidence in constructing classifications reflects the schools' different ontological commitments. For cladists, only monophyletic taxa are "natural" taxa (Wiley 1981). Because only synapomorphies serve as evidence for monophyletic taxa, cladists use only that type of character in constructing classifications. Evolutionary taxonomists believe that both monophyletic and paraphyletic taxa are natural, so they use symplesiomorphies as well as synapormorphies. Still, evolutionary taxonomists preclude the positing of polyphyletic taxa. They argue that polyphyletic taxa are not the result of common ancestry and thus are not natural. Though evolutionary taxonomists are ontologically more conservative than pheneticists, they are too liberal for cladists. Evolutionary taxonomists allow the positing of paraphyletic taxa whereas cladists, for the most part, forbid it. A paraphyletic taxon does not contain all the descendants of its common ancestor – it is an incomplete branch. This disagreement over the status of paraphyletic taxa marks a philosophical difference between cladism and evolutionary taxonomy. Evolutionary taxonomists allow that some qualitative features, namely, the adaptive complexes of grades, are important in classifying organisms. Cladists believe that only similarities reflecting common ancestry count.

Though evolutionary taxonomy and process cladism differ significantly, they share some methodological principles that distinguish them from phenetics and pattern cladism. Both evolutionary taxonomists and process cladists believe that process assumptions concerning evolution are essential for constructing classifications. Pheneticists and pattern cladists, on the other hand, argue that such assumptions are not essential. Moreover, they contend that the principles of taxonomy should be void of such assumptions.

I mention these three strands of contention in part because they nicely summarize the differences among the four major schools, but also because they will resurface in later chapters. I should emphasize, however, that the focus of this book is not the debates among the various schools but several particularly philosophical issues in biological taxonomy, namely, the historical turn in biological classification, the possibility of taxonomic pluralism, and the future of biological nomenclature. Still, disagreements among the schools will arise. In the discussion of the historical turn (Chapter 3), what sorts of similarity serve as evidence for classification is an important issue, as is the question of whether process assumptions should be used in constructing classifications. In the discussion of pluralism (Part II), a stand will be taken on what sorts of taxa should be classified; to do otherwise allows a form of pluralism where "anything goes." The discussion of species pluralism (Chapter 4) raises the question of whether paraphyletic taxa exist and whether they should be classified. The debates among the four schools are not the focus of this book, but they are an undercurrent.

2.5 SPECIES CONCEPTS

Earlier sections briefly introduced the four major schools of biological taxonomy. In this section we turn to a more limited topic but one that is exceedingly important to biologists: the nature of species. Species are the base units in the Linnaean hierarchy. Many biologists believe that the importance of species goes beyond their role in the Linnaean hierarchy. Species are seen as the fundamental units of evolution, what might be described as "the engines of evolution" (Eldredge and Cracraft 1980, Wiley 1981, Mayr 1982, Ghiselin 1987). All other types of taxa, genera, families, and so on, are passive aggregates composed of actively evolving species. For these biologists, a key to understanding evolution is understanding the nature of species. Many believe that this is why Darwin titled his landmark book *The Origin of Species*.[4]

Before introducing various accounts of species, it is important to distinguish two types of entities: *species taxa* and the *species category*. Species taxa are groups of organisms; *Homo sapiens* and *Drosophila melanogaster* (fruitflies) are examples of species taxa. The species category is a much more inclusive entity: it is the class of all species taxa. In this section, we are interested in the definitions biologists provide for the species category.[5] Biologists refer to definitions of the species category as "species concepts." A species concept tells us what species taxa have in common so that they are members of the species category.

In the previous sections of this chapter, we discussed no fewer than four general schools of biological taxonomy. When we turn to contemporary species concepts, our options multiply. The current biological literature contains well over a dozen species concepts. Disagreement over the nature of species is far from new. Biologists have offered different definitions of the species category for hundreds of years. For example, in a letter to Joseph Hooker, Darwin writes:

> It is really laughable to see what different ideas are prominent in various naturalists' minds, when they speak of "species"; in some, resemblance is everything and descent of little weight – in some resemblance seems to go for nothing, and Creation the reigning idea – in some, descent is the key, – in some sterility an unfailing test, with others it is not worth a farthing (F. Darwin 1877, volume 2, 88).

As we shall see, the issue is still far from settled. Our survey of contemporary species concepts includes seven prominent concepts. We begin with the best known and most influential species concept of the last fifty years, Mayr's biological species concept.

The Biological Species Concept

Mayr provides several versions of the biological species concept. His most widely accepted one is the following: "Species are groups of natural populations that are reproductively isolated from other such groups" (Mayr 1970, 12). Mayr first presented the biological species concept in his essay, "Speciation Phenomena in Birds" (1940). Dobzhansky introduced an earlier version in his *Genetics and the Origin of Species* (1937). The history of defining the species category in terms of reproductive criteria goes back much further. Consider Buffon's definition:

> [W]e should regard two animals as belonging to the same species if, by means of copulation, they can perpetuate themselves and preserve the likeness of the species; and we should regard them as belonging to different species if they are incapable of producing progeny by the same means (Buffon 1749, 10; quoted in Lovejoy 1959, 93).

Other biologists who used reproductive criteria for defining the species category include John Ray (1686), Cuvier (1815), and Jordon (1905). (See Mayr 1982, 270ff., and Dobzhansky 1970, 353ff., for discussions of the history of the biological species concept.)

According to Mayr's biological species concept, a species is a group of organisms that interbreed and produce fertile offspring. That group may consist of a single local population or a number of populations connected by at least occasional interbreeding. The members of different species, on the other hand, are prevented from producing fertile offspring. The organisms in different species are separated by "reproductive isolating mechanisms" that either prevent interbreeding from occurring ("prezygotic isolating mechanisms") or prevent the production of fertile offspring if interbreeding does occur ("postzygotic isolating mechanisms"). Examples of prezygotic isolating mechanisms are incompatibility between male and female gametes and temporally distinct mating seasons. Examples of postzygotic mechanisms include the production of inviable or sterile hybrids.

The biological species concept is explicitly defined in terms of isolating mechanisms, but the concept is largely motivated by the idea that species taxa form a unique and important type of genetic system. As supporters of the biological species concept argue, "a species is the most inclusive Mendelian population" (Dobzhansky 1970, 357); it is "a field for gene recombination" (Carson 1957). In other words, the biological species concept places species taxa within the purview of population genetics.[6] A biological species is a distinct gene pool whose genetic material is recombined in every generation. The boundaries of that pool are determined by which organisms can combine genetic material through interbreeding and produce fertile offspring. The movement of genetic material among the members of a species is called "gene flow."

Isolating mechanisms mark the boundaries of individual species taxa, but more important, they are essential for the continued existence of species taxa. Mayr writes that "every species is an ecological experiment, an attempt to occupy a new niche" (1970, 357). A species is successful only if its organisms are sufficiently adapted to their environment. And a species will continue to be successful only if it is protected from the incursion of genes that would cause its organisms to become maladaptive. Thus, "the major meaning of reproductive isolation is that it provides protection for a genotype adapted for utilization of a specific niche" (Mayr 1982, 275). Species are stable gene pools, protected from the incursion of foreign genetic material by isolating mechanisms. A species' stability is also supported by interbreeding. A mutation that enhances the adaptiveness of an organism may take hold in a population of a species, causing that population to diverge from the rest of the species. Interbreeding between that population and the rest of the species may spread the trait and preserve the unity of the species. On the biological species concept, species are genetic fortresses, protected by isolating mechanisms and held together by interbreeding.

Along with the biological species concept, Mayr argues for a closely related model of speciation. Speciation models come in various forms; two main types are *allopatric speciation* and *sympatric speciation*. In allopatric speciation, geographically separated populations evolve independently, resulting in speciation. In sympatric speciation, speciation occurs without the geographic separation of populations. Mayr advocates a particular form of allopatric speciation, sometimes called *peripatric* speciation. In this form of allopatric speciation, a small population or group of migrants is geographically separated from the rest of the species. The isolated population and the main body of the species no longer exchange genetic material through gene flow. The isolated population may be exposed to different ecological factors, causing it to diverge genetically from the main body of the species. Furthermore, the isolated population may be exposed to mutations and instances of genetic drift (random genetic sampling) different from those in the main body of the species, providing further chances for divergence. If the isolated population were connected by gene flow to the rest of the species, such divergence would be opposed by the influx of genetic material from the rest of the species. But because it is not, the isolated population is vulnerable to genetic change. Consequently, the population may undergo a genetic revolution – a drastic genetic restructuring required to survive in its new ecological and genetic environments.

In Mayr's model of speciation, two factors must be met for speciation to occur. First, an isolated population must be genetically restructured so that it survives in its new environment. Second, it must develop reproductive isolating mechanisms to protect its adapted gene pool from the incursion of genetic material from other species. This raises an important question: How do such isolating mechanisms develop? According to Mayr, reproductive isolating mechanisms "arise as an incidental by-product of genetic divergence in isolated populations" (1970, 327). There is no selection for reproductive isolating mechanisms because such populations are already isolated. Isolating mechanisms are merely by-products of organisms becoming adapted to their new ecological and genetic environments.

The link between the biological species concept and Mayr's model of allopatric speciation is not hard to see. The concept states that "species are groups of interbreeding natural populations that are reproductively isolated from other such groups." In Mayr's model of speciation, a speciation event is initiated only when interbreeding between a geographically isolated population and a species' main body is interrupted. Furthermore, speciation is complete only when an incipient species becomes reproductively isolated from its ancestor.

Mayr does not put much stock in the other main model of speciation, sympatric speciation. In that model, speciation can occur even though the populations involved are not geographically isolated. Mayr believes that sympatric speciation rarely occurs in nature. His reasoning is this:

> The essential component of speciation, that of the genetic repatterning
> of populations, can take place only if these populations are temporarily
> protected from the disturbing inflow of alien genes. Such protection can

best be provided by extrinsic factors, namely spatial isolation. It appears that such spatial isolation is normally effected by geographical barriers (1970, 277).

So Mayr's case against sympatric speciation is an empirical one: isolation sufficient for a genetic revolution rarely occurs unless a population is geographically isolated.

Still, Mayr allows that one form of sympatric speciation does occur, namely, speciation initiated by polyploidy. Polyploidy occurs when offspring have more chromosomes than their parents. Suppose that the typical diploid chromosome number found in a species is 12. Offspring that have chromosome numbers greater than 12 and in multiples of 6 (18, 24, 30 . . .) are polyploids. Because polyploids have a different chromosome number from their parental species, mating between polyploids and members of the parental species tend to produce inviable or infertile offspring. As a result, polyploids acquire reproductive isolating mechanisms without the need for geographic isolation. Polyploids can satisfy the biological species concept's requirement of reproductive isolation through sympatric speciation. Polyploidy is rare in animals, but is widespread in plant species (Mayr 1970, 254–5).

The Recognition Species Concept

In the last thirty years, numerous objections have been raised to the biological species concept. Those objections often serve as the basis for alternative species concepts. One such concept is Paterson's (1985) recognition species concept. Paterson's objection to the biological species concept centers on Mayr's account of geographic (allopatric) speciation.

Paterson agrees with much of Mayr's model of speciation. He agrees that speciation occurs only when the members of a species acquire reproductive isolating mechanisms. And he concurs with Mayr that the acquisition of such mechanisms usually occurs during geographic isolation. However, Paterson disagrees with Mayr concerning what process causes the existence of such mechanisms. According to Mayr, the rise of reproductive isolating mechanisms is merely an accidental by-product of a population's genetic reconstruction. Paterson, on the other hand, suggests that there are forces directly responsible for the existence of such mechanisms. We just need to view isolating mechanisms differently. Rather than being treated as mechanisms that prevent reproduction among species, they should be recognized as mechanisms that promote successful interbreeding within species. Paterson argues that within isolated populations there is selection for organisms that recognize potential mates with a similar way of life. That selection causes members of isolated populations to acquire what Paterson calls "specific mate recognition systems." The characteristics of such systems are diverse: many birds use calls and songs to recognize their mates; moths, bees, and wasps use chemical signals; and most flowering plants have discerning stigmas.

Both Paterson's and Mayr's models of speciation require that the organisms of a species obtain mechanisms for discriminating among prospective mates. But according to Paterson, only his model provides an adaptational explanation for why organisms develop such mechanisms. As a result, Paterson contends that his model of speciation should be accepted in place of Mayr's. Moreover, Paterson believes that Mayr's failing to provide an adaptational explanation indicates a weakness in his species concept. Recall that Mayr defines species as "groups of interbreeding natural populations that are *reproductively isolated* from other such groups" (italics added). Because there is no adaptational explanation for the existence of purely reproductive isolating mechanisms, Paterson argues that the second part of Mayr's definition should be eliminated. Instead, Paterson proposes that a species is "that most inclusive population of individual biparental organisms that share a common fertilization system" (1985[1992], 149). Paterson calls his species concept "the recognition concept of species" because of the importance he attaches to mate recognition systems.

Despite their differences, Mayr's and Paterson's definitions of the species category agree in important ways. For both, a species is a group of interbreeding organisms, and the organisms of a species cannot successfully interbreed with organisms of other species. Both of their definitions concur with a central tenet of the evolutionary synthesis, that species are closed gene pools in which recombination occurs. Where Mayr and Patterson disagree is how individual species become distinct gene pools, and whether the mechanisms that maintain such pools should be viewed as reproductive isolating mechanisms or mate recognition systems.

The Phenetic Species Concept

The phenetic species concept is an application of pheneticism to the species category. As we saw in Section 2.2, pheneticists advocate a theory-free, empiricist approach to classification for several reasons. One is their concern that a theoretically motivated approach to classification causes biologists to classify certain patterns in nature while ignoring others. Another motivation is the desire to have classifications that are more lasting than the latest biological theory. Pheneticists suggest that our classifications would have more permanence if they were devoid of theoretical assumptions. These sorts of reasons cause pheneticists to reject the biological species concept and offer their own species concept. Sokal and Crovello (1970[1992], 50–1) write:

> [I]f we assume a priori that all organisms can be put into some biological species, then we of necessity concentrate on finding such classes. . . . The emphasis [should] be on unbiased description of the variety of evolutionary patterns that actually exist among organisms in nature, and of the types of processes bringing about the different varieties of population structure. We believe that in the long run this approach would lead to greater and newer insights into the mechanisms of evolution.

Accordingly, Sokal and Crovello offer an overtly phenetic species concept: species are simply those organisms that have the most overall similarity. Such similarities include macrosimilarities – for example, morphological, physiological, behavioral, and ecological similarities, as well as microsimilarities, such as DNA homologies and similarities in amino acid sequences. In principle, the phenetic methodology for arriving at species taxa is straightforward. A number of organisms are assembled; their similarities and dissimilarities are recorded. Given enough data, and the use of statistical methods, a pattern should develop with certain groups having the most overall similarity among their members. Those groups are species taxa. The whole procedure is supposed to take place without the use of any assumptions from evolutionary theory: just observe and calculate which clusters of organisms have the most overall similarity.

A further reason that Sokal and Crovello prefer the phenetic species concept is their belief that biologists, even supporters of the biological species concept, use phenetic methods when identifying species. They illustrate this point by highlighting the steps required for identifying a group as a biological species. The members of a biological species must be physiologically capable of interbreeding – that is, they must be interfertile. But given the limited time and resources available to biologists, tests for interfertility are restricted to small samples of the group in question. Whether the entire group forms an interfertile population is settled by the phenotypic similarity between the sampled and unsampled organisms in the group. Sokal and Crovello (1970[1992], 45–6) conclude, "we are left with what is essentially a phenetic criterion of homogeneous groups that show definite aspects of geographic connectedness and in which we have any evidence at all on interbreeding in only a minuscule proportion of cases." The inference they draw is that proponents of the biological species concept are, for the most part, using the phenetic species concept.

The Ecological Species Concept

Mayr's and Paterson's species concepts, though different in important ways, fall under the general heading of an interbreeding approach to species. Pheneticists object to the interbreeding approach to species on methodological grounds. Other biologists question the interbreeding approach's empirical adequacy. Recall that proponents of the interbreeding approach believe that species are stable taxonomic units because the members of a species can successfully interbreed. Ehrlich and Raven (1969), Van Valen (1976[1992]), and Andersson (1990) disagree with this empirical claim. They contend that the stability of a species is primarily due to environmental forces rather than interbreeding. They argue that an ecological species concept should replace concepts based on interbreeding.

In the first leg of their argument, proponents of the ecological approach suggest that interbreeding is neither necessary nor sufficient for the existence of species taxa. It is not necessary, they maintain, because many species consist of organisms that do not exchange genetic material through interbreeding. Asexual organisms, they argue, form species, yet asexual organisms do not interbreed. In addition, many species of sexual organisms consist of geographically separated populations that exchange little if any genetic material. For example, the sand crab *Emerita analoga* consists of disjunct populations located in the northern and southern hemispheres, yet it persists as a species (Ehrlich and Raven 1969). Turning to the sufficiency of interbreeding for maintaining species taxa, proponents of the ecological species concept cite the existence of multispecies. A multispecies is a group of species that remain distinct taxa despite their organisms hybridizing and producing fertile offspring. A classic example is North American oaks: the Canadian species *Quercus macrapa* and *Q. bicolor* often interbreed and produce fertile offspring; nevertheless, those species remain distinct taxonomic units (Van Valen 1976[1992]).

Proponents of the ecological species concept suggest that natural selection is the primary force preserving species. Species consisting of geographically isolated populations or those consisting of asexual organisms are unified by their members being exposed to similar sets of selection forces (what some authors refer to as "common selection regimes"). Van Valen (1976[1992], 70) incorporates this suggestion in an ecological species concept: "A species is a lineage . . . which occupies an adaptive zone minimally different from that of any other lineage in its range and evolves separately from all lineages outside its range." What is an "adaptive zone"? Van Valen (1976[1992], 70) describes it as

> some part of the resource space together with whatever predation and parasitism occurs on the group considered. It is part of the environment, as distinct from the way of life of a taxon that may occupy it, and exists independently of any inhabitants it may have.

Adaptive zones are what biologists commonly call "niches." So each species, according to Van Valen, occupies its own niche, and the selection forces of that niche preserve the species as a distinct taxon.

It should be noted that according to the ecological species concept, membership in a species does not merely turn on whether two organisms occupy the same adaptive zone. The ecological species concept places two constraints on species: the members of species must live in a similar adaptive zone, and those members must be part of a single lineage. Both requirements are necessary, but neither is sufficient for species membership. Consequently, two organisms living in different parts of the world but exposed to similar selection forces may or may not be conspecific according to the ecological species concept. It all depends on whether those organisms are also parts of the same lineage. The ecological species concept, despite its name, is a genealogically based concept: it assumes that ecological forces segment the tree of life into base lineages; and it assumes that ecological forces maintain those lineages as distinct species.

The Evolutionary Species
Concept

The biological, recognition, and ecological species concepts each highlight a single type of process as responsible for the existence of species.

88

The evolutionary species concept allows that a variety of processes may be responsible for species. The evolutionary species concept was first introduced by Simpson (1961) and later amplified by Wiley (1978). Consider Wiley's (1978[1992], 80) version: "A species is a single lineage of ancestral descendant populations of organisms which maintains its identity from other such lineages and which has its own evolutionary tendencies and historical fate." Species taxa are lineages with their own evolutionary tendencies despite the different types of processes responsible for those tendencies. In many species, interbreeding and reproductive isolation are responsible for those tendencies. For example, two local populations may have similar evolutionary tendencies due to the extensive gene flow among their members. However, species consisting of geographically isolated populations or asexual populations owe their unique tendencies to other sorts of processes. The unity of such species is the result of their organisms' being exposed to similar selection pressures or having similar epigenetic or developmental constraints.

One question often asked about the evolutionary species concept is what is meant by the notion of unique "evolutionary tendencies and historical fate" (see Templeton [1989(1992)]). Species are supposed to have them, but what are they? When Wiley (1981, 25ff.) explains how species have their own evolutionary tendencies and historical fates, he is quick to point out that such tendencies and fates do not require phenotypic similarity among the members of a species. Variation at a time and variation over time are still the norm. Instead, Wiley suggests that the members of a species are exposed to similar types of processes that cause those members to react in relatively similar ways to genetic changes. For example, gene flow may cause the spread of an adaptive gene to future generations of a species. Conversely, homeostatic genotypes among the members of a species may prevent the occurrence of new traits within the local populations of a species. That, in turn, prevents populations from diverging from the rest of the species. As another example, stabilizing selection may prevent divergence within a species of asexual organisms. Again, exposure to such mechanisms causes the members of a species to react in relatively similar ways to genetic changes. That causes a species to have its unique "evolutionary tendencies and historical fate."

Simpson first introduced the evolutionary species concept to overcome what he saw as the overly restrictive nature of the biological species concept (1961, 153ff.). The biological species concept does not apply to asexual organisms, nor does it apply to extinct species whose fossils lack information concerning their ability to interbreed. The evolutionary species concept, suggests Simpson, is applicable in such cases because it highlights a more fundamental aspect of species than the biological species concept. Asexual species, extinct species, and good biological species are all lineages with their "own evolutionary tendencies." Thus, all form species according to the evolutionary species concept.

Phylogenetic Species Concepts

A number of recent species concepts have been classified under the heading of the "phylogenetic species concept." Each tries to provide an account of species that falls within the tenets of cladism. Just as the school of cladism is divided into two camps, process cladism and pattern cladism, so are phylogenetic species concepts. Some phylogenetic species concepts are developed with an eye toward processes (Mishler and Donoghue 1982, Mishler and Brandon 1987, Ridley 1989, Baum and Shaw 1995). Other phylogenetic species concepts focus on patterns (Nelson and Platnick 1981, Cracraft 1983, Nixon and Wheeler 1990, Davis and Nixon 1992).[7] Let us begin with a process-oriented phylogenetic species concept.

Hennig, the father of cladism, reserved the concept of monophyly for higher taxa. He did not apply it to species taxa. Mishler and Brandon (1987) want to bring the species category in line with cladism and require that species taxa form monophyletic groups. The biological species concept cannot meet this requirement because some biological species are paraphyletic. Consider the freshwater fish group *Xiphophorus*, which consists of a number of populations (Figure 2.12). The members of C and F successfully interbreed and are reproductively isolated from the members of the other populations. According to Rosen (1979, 275–9), C + F forms a single biological species. However, C + F does not contain all the descendants of the common ancestor X. So C + F is a paraphyletic biological species. (More examples of paraphyletic biological species are given in Chapter 4.) Given the existence of such species, Mishler and Brandon argue that the biological species concept should be replaced with a concept requiring that species form monophyletic groups.[8] They offer the following phylogenetic species concept:

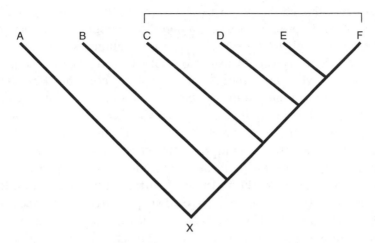

Figure 2.12   A cladogram of the freshwater group *Xiphophorus*. C + F is a biological species as well as a paraphyletic taxon. Adapted from Rosen 1979, 276.

> A species is the least inclusive taxon recognized in a classification, into
> which organisms are grouped because of degree of monophyly (usually,
> but not restricted to, the presence of synapomorphies), that is ranked as
> a species because it is the smallest "important" lineage deemed worthy
> of formal recognition, where "important" refers to the action of those
> processes that are dominant in producing and maintaining lineages in
> a particular case (1987, 46).

This concept places two conditions on species taxa – what Mishler
and Brandon refer to as a "grouping criterion" and a "ranking criterion." Their ranking criterion requires that all species be monophyletic
taxa. This brings species taxa in line with the general cladistic prescription that all taxa be monophyletic. Since all taxa are monophyletic,
according to cladists, a ranking criterion is needed to distinguish species
taxa from genera and other higher taxa. Mishler and Brandon allow
that various types of processes may cause taxa to be species. Some taxa
are ranked as species because of interbreeding among their members.
Other taxa are ranked as species because of the similar ecological or
developmental forces affecting their organisms. Mishler and Brandon's
phylogenetic species concept is pluralistic concerning which processes
cause a taxon to be a species, but it is monistic in that all species taxa
be monophyletic.[9]

Pattern cladists, as should come as no surprise, prefer species concepts void of any reference to processes. Pattern cladists are in the busi-

ness of recognizing the patterns of nature. As we saw in Section 2.4, pattern cladists offer various arguments for adopting a theory-free, cladistic approach to classification. Those arguments apply to their species concepts as well. Consider one of Cracraft's (1983[1992], 103) reasons for advocating a pattern cladist species concept: "by defining species in terms of the resulting *pattern*, it allows us to investigate these processes, unbiased by a species concept that is derived from our pre-conceptions of those processes." Pattern cladists, as pheneticists, believe that theoretical assumptions about processes cause biologists to misrepresent nature's patterns. Accordingly, they advocate species concepts that depend only on theory-neutral observations of pattern. Here are two examples of pattern cladist species concepts. According to Nelson and Platnick (1981, 12), "species are simply the smallest detected samples of self-perpetuating organisms that have unique sets of characters." More recently, Davis and Nixon (1992, 427) suggest that species are "the smallest aggregation of populations (sexual) or lineages (asexual) diagnosable by a unique combination of character states in comparable individuals."

Notice that these species concepts are not completely void of any reference to process. Both assume that the members of a species are connected by heredity relations. Nevertheless, to determine which genealogical chunks are species, these concepts rely on phenotypic or genotypic similarities among the members of a species. For example, Nelson and Platnick's species concept relies on "unique sets of characters" for identifying which lineages are species taxa. This is in contrast to process species concepts that rely on the occurrence of a common set of processes for identifying species. The pattern cladist shift from process criteria to phenotypic similarity, however, does not imply that pattern cladist species concepts reduce to the phenetic species concept. Pheneticists form taxa on the basis of overall similarity. Pattern cladists, as we saw in Section 2.4, are not advocating that. Only derived character states (or synapomorphies) serve as evidence for pattern cladists.

The last part of the chapter is a quick introduction to seven prominent species concepts. This introduction is far from exhaustive given that the recent biological literature contains well over a dozen species concepts (just review back issues of *Systematic Biology* [formerly *Systematic Zoology*] and *Systematic Botany* for the last twenty years). We have seen why proponents of different concepts prefer one concept over another. No attempt, however, has been made to resolve these debates or to say which concept(s) should be preferred. In Chapters 4 and 5, we will discuss the merits and problems of various species concepts. We will also consider the philosophical question of whether a plurality of different species concepts should be accepted. But before considering such issues, we need to discuss a more fundamental issue: the historical turn in biological taxonomy.

3

History and Classification

Chapter 1 outlined three main approaches to classification: essentialism, cluster analysis, and the historical approach. Prior to Darwin's
work, the prominent view among biologists was that essentialism and
cluster analysis are the proper methods for sorting organisms into
species. Since then, the historical approach has become the dominant
method. Nevertheless, the shift from essentialism and cluster analysis
to the historical approach has been controversial. Those biologists who
write on the theoretical aspects of biological classification almost
universally concur with the shift to the historical approach. Philosophers, however, remain divided. Hull (1976, 1978), Sober (1980, 1984a),
Rosenberg (1985a), Williams (1985), and Ereshefsky (1991a) champion
the historical approach to biological classification. Kitts and Kitts
(1979), Dupré (1981, 1993), Kitcher (1984a, 1984b), and Ruse (1987)
favor more qualitative approaches. The first half of this chapter takes
up the debate over which approach is appropriate for biological
taxonomy. Sections 3.1 and 3.2 outline problems in applying essentialism and cluster analysis to biological taxonomy. Section 3.3
shows why the historical approach is the proper one for biological
classification.

Closely associated with the historical approach is the now infamous
"species are individuals" thesis. Unfortunately the term "individual"
has taken on several meanings in the debate over whether species are
individuals. For some authors, an entity is an individual if it is spatiotemporally continuous. For others, individuality requires more than
mere spatiotemporal continuity. Disagreement over the meaning of
individuality has led to undue confusion. Section 3.4 enters the debate
first by distinguishing various notions of individuality and then by
examining which of those notions apply to species.

The proposal that species are individuals has caused quite a stir in biology and the philosophy of biology. Part of that interest is due to the foundational nature of the issue – nothing less than the question of how biological taxonomy should be conducted. But part of that interest is due to what many see as the wider implications of the historical nature of biological taxa. Some believe that the individuality of species has important implications for how evolution works – for example, whether the tempo of evolution is gradual or punctuated and whether species selection occurs (Eldredge 1985, Cracraft 1987).[1] Others see more philosophical implications of the historical nature of taxa. They believe that the historical nature of taxa renders evolutionary theory a different sort of theory from those found in physics or chemistry (Smart 1968, Gould 1986, Mayr 1988, O'Hare 1988). Section 3.5 shows that the genealogical nature of taxa does not create a methodological divide between evolutionary biology and the physical sciences.

3.1 THE DEATH OF ESSENTIALISM

Recall the three main tenets of essentialism (Section 1.1). (1) All and only the members of a kind share a common set of traits that is the real essence of that kind. (2) That essence makes an entity a particular kind of thing. (3) Knowing an entity's essence is crucial for explaining the other properties typically associated with an entity of that kind. Prior to the acceptance of evolutionary theory, essentialism was the standard mode of classification in biological taxonomy. Such biologists as John Ray, Maupertuis, Bonnet, Linneaus, Buffon, and Lamarck believed that the proper way to sort organisms into species taxa is by their species-specific essences (Hull 1965; Sober 1980; Mayr 1982, 256ff.). However, aspirations of essentialism in biological taxonomy have floundered on theory and messy empirical details.

To begin, Hull (1965 [1992], 202–4) has observed that essentialism does not sit well with early notions of evolution. Both Lamarck and Darwin were evolutionary gradualists. They believed that change within a species is a slow and constant process. Speciation is merely the accumulation of small changes whereby one species slowly evolves into another. As Hull (1965 [1992], 203) points out, the joint adoption of gradualism and essentialism leads to a problem:

> If species evolved so gradually, they cannot be delimited by means of a
> single property or set of properties. If species can't be so delineated, then
> species names can't be defined in the classic manner. If species names
> can't be defined in the classical manner, then they can't be defined at all.
> If they can't be defined at all, then species can't be real.

The boundaries of species, in other words, are vague in the same sense
that the boundary between bald and not bald is vague. No precise
number of hairs marks the boundary between bald and not bald. Similarly, no genetic or phenotypic trait marks the boundary from one
species to the next. Thus a supporter of evolutionary gradualism and
essentialism must give up one of three things: essentialism, gradualism,
or the existence of species taxa. As Hull indicates, the above argument
led Lamarck to reject the existence of species taxa but to retain his
beliefs concerning essentialism and gradualism.

A quick look at recent biology might cause one to wonder if the
problem here is not the existence of species taxa but gradualism. Recall
the peripatric model of speciation discussed in Section 2.5. A small population becomes geographically isolated from the main body of its
species and is exposed to new selection pressures. Because its gene pool
is both limited and isolated, that population responds to those pressures by undergoing a genetic revolution. That revolution results in a
new species when the isolated population acquires a set of stable and
well-adapted genotypes and its members cannot successfully interbreed with members of its parental species. This account of speciation
is the cornerstone of the theory of punctuated equilibria (Eldredge and
Gould 1972). According to Eldredge and Gould, most evolutionary
change occurs during the genetic revolutions of geographic speciation.
Between such speciation events, little if any significant evolutionary
change occurs. The theory of punctuated equilibria stands in stark contrast to evolutionary gradualism: the theory of punctuated equilibria
posits quick and infrequent bursts of evolution; evolutionary gradualism posits slow and constant change. So, one way to lessen the tension
between essentialism and evolutionary gradualism is to reject gradualism and opt for the theory of punctuated equilibria.

Still, this maneuver does not save essentialism. Even on a nongrad-
ualist account of evolution, speciation is an extended process. Peripatric speciation involves a number of processes, including the
introduction of new traits via mutation and recombination, selection
for new genotypes, and, in the case of sexual species, the development
of reproductive isolating mechanisms. These processes do not occur

overnight but often span hundreds and thousands of generations (Futuyma 1986, 245). Such a picture of evolution departs from gradualism in that significant change within a species occurs at speciation and not throughout its life. Even so, geographic speciation is often a gradual event, and Lamarck's concern that we cannot draw a precise boundary between species remains. So despite the rejection of gradualism, species essentialism may still be undermined by the vagueness of speciation events.

Sober (1980 [1992], 252ff.), however, argues that the vagueness of speciation events is not fatal for essentialism. He suggests that essentialism is consistent with vague boundaries so long as essences are correspondingly vague. Consider an example. We observe discrete jumps on the periodic table between the atomic numbers of chemical elements. Nevertheless, the boundary between types of atoms is vague. At what point, for instance, does an atom cease to be a nitrogen atom and become an oxygen atom? According to standard chemistry texts, it is when an atom absorbs an alpha particle and gives off a proton. But that process takes time, thus creating a vague boundary between nitrogen and oxygen. Sober contends that though nitrogen may consist of borderline cases, that does not defeat essentialism if those cases coincide with the vagueness of having the atomic number 14. The same reasoning applies to species. Species boundaries may be vague, but that does not defeat species essentialism so long as the vagueness of those boundaries coincides with the vagueness of species essences.

Does Sober's addendum to essentialism save species essentialism from the fact of multigenerational speciation? I am doubtful. A fundamental tenet of essentialism is that each entity has a determinate real essence that crucially explains many of that entity's other characteristics. Sober's revision of essentialism is inconsistent with that tenet. Consider the appeal Sober (1993, 148) makes for vague essentialism:

> I suspect that no scientific concept is *absolutely* precise; that is, for every concept, a situation can be described in which that concept's application is indeterminate. Essentialism can tolerate imprecisions of this sort.

Oddly, the first sentence in this quotation captures the heart of Locke's vagueness argument *against* essentialism (see Section 1.1). According to Locke, if kinds have vague essences, then their individual members can have vague essences. But an entity lacking a determinate essence is no entity at all because an entity can exist only as a particular kind of thing. More practically, if an entity lacks a specific essence, then essentialists have nothing to appeal to in explaining its nature. If Sober is right that no scientific concept is absolutely precise and consequently some entities have indeterminate essences, then instead of shoring up essentialism he has furthered the case against it.

Returning to species, we see that the existence of multigenerational speciation spells trouble for species essentialism. Essentialism requires sharp boundaries and precise essences; yet speciation is often a vague process in which no set of traits marks a species' boundary. Consequently, species essentialism is at odds with how speciation often occurs.

The above is just one of many reasons for rejecting species essentialism. Another is that biologists have been hard-pressed to find traits that occur among all and only the members of a particular species (Hull 1965 [1992], 205). Contemporary evolutionary theory explains why. A number of forces counter the universality and uniqueness of a biological trait within a species. Suppose a genetically based trait were found in all the members of a species. Mutation, random drift, or recombination can cause that trait to disappear in a future member of that species. All it takes is the disappearance of a trait in a single member of a species to show that it is not essential. On the other hand, suppose all the members of the species have a common trait. In order for it to qualify as an essential property that trait must also be unique to the members of the species. Yet organisms in different species often have common characteristics. Again evolutionary biology explains why. Different species frequently live in similar habitats. This causes parallel selection for the same traits in the organisms of various species. Furthermore, the organisms of different species inherit similar genes from a common ancestor. Such genetic material is another cause of similar traits in the members of various species.

Despite the above arguments, Kitcher (1984a, 1984b) has argued that species may have necessary traits – traits that each member of a species must have to be a member of that species. He suggests two types of necessary traits. The first type of trait is one whose absence causes the existence of inviable offspring. Sober (1984a) has responded that the lack of such a trait does not signal a species boundary. Within selection theory, an inviable zygote is counted as a mortality within its parents' species, not a member of a different species. This accords with our intuitions. An aborted human fetus is an inviable zygote, yet we still take it to be a member of the species *Homo sapiens* (questions about persons and souls notwithstanding). Although Kitcher has suggested a property that may be necessary for having viable offspring, he has not suggested a property necessary for membership within a species. Essentialism requires the latter, not the former.

The second necessary property Kitcher offers is one whose absence would cause viable offspring to belong to a different species from their parents. Kitcher has in mind speciation through polyploidy, where offspring have twice the number of chromosomes of their parents. For example, suppose the members of a species typically have 12 chromosomes. Offspring with 24 chromosomes are polyploids. Kitcher suggests that all members of the original species must have fewer than 24 whole chromosomes; any more and an organism belongs to a different species. Again this suggestion runs into trouble. Polyploidy is not considered a sufficient condition for speciation. According to Briggs and Walters (1984, 242, 251), a group of polyploids does not form a new species unless that group also becomes reproductively isolated and competitively successful. The occurrence of polyploidy by itself does not signal a species boundary; we must also consider what happens to those organisms after the initial increase in chromosomes. Once again we see that speciation is not complete at the first instance of genetic change or geographic isolation.

Kitcher has not come up with an obvious necessary trait for species membership, but even if we were more sanguine about Kitcher's examples, they would not establish species essentialism. In both cases, the traits he offers are found in all the members of a species; however, species essences must be traits found in all and *only* the members of a species. Neither of the traits Kitcher has suggested plausibly occur in only the members of a particular species. Complications that lead to inviable zygotes are often not species specific. Furthermore, many species have common chromosome numbers. Instead of supporting species essentialism, Kitcher's examples nicely illustrate the lengths one might go to find an instance of a species essence and still be unsuccessful.

Despite the above arguments, one might wonder if the human genome project is on the way to discovering the genetic essence of *Homo sapiens*. Again, I am pessimistic. Suppose the human genome project produced a model of an entire human genotype. That model would not provide the essence of *Homo sapiens*. As high school genetics teaches us, different humans have different genotypes. Nevertheless, the project has revealed that particular genes are important in leading, by society's standards, a healthy, normal life. Does the aggregate of such genes constitute the essence of *Homo sapiens*? The answer is no. Having health problems, and genes that give rise to them, does not make one any less a member of *Homo sapiens*. Alternatively, one might try to equate the essence of *Homo sapiens* with the genotype, or some aspect of it, that most humans have. But that won't save essentialism either. Essentialism requires a set of traits that all, not merely most, humans have. Perhaps essentialism should instead highlight the genes that make a human fit in his or her environment. That option fails because it would imply that if I were marooned on Mars without a space suit I would not be a *Homo sapiens*. A final suggestion for the essence of *Homo sapiens* is the suite of genes that makes an organism fit in the normal environment of *Homo sapiens*. Even if we ignore the problem of whether species have normal environments, this suggestion will not work. If the essence of *Homo sapiens* is the suite of genes that allows an organism to be fit in the normal environment of *Homo sapiens*, then many cats, dogs, and cockroaches are *Homo sapiens*.

Thus far we have seen that nature itself conspires against species essentialism. Some biological processes work to break the universality of a trait in a species while others confound the uniqueness of a trait. Along with these empirical reasons for rejecting species essentialism are more theoretical ones. One motivation for positing essences is to explain the behavior of individuals and the groups they comprise. But the mechanisms posited by essentialists to explain life's diversity have no home in evolutionary theory (Mayr 1959, Sober 1980). Recall Aristotle's dispositional account of essences (Section 1.1). The organisms of a species have a common natural state. An organism's essence is its disposition to achieve its natural state provided that nothing interferes with its ontogenetic development. However, interfering forces usually affect the development of organisms, and as a result the members of a species vary in their traits. The dispositional account allows Aristotle, and such pre-evolutionary biologists as John Ray, Maupertuis, and Bonnet, to preserve species essentialism in the face of observed variation (Sober 1980).

As Sober (1980) shows, this dispositional account of species essentialism has no basis in contemporary evolutionary theory. Suppose we give the following genetic reading of the dispositional account. The members of a species share similar genotypes, and those genotypes cause the natural phenotype for the members of that species, provided no interfering forces occur. This version of the dispositional account is inconsistent with contemporary genetics. In population genetic texts, one finds a "norm of reaction" that graphs the different phenotypes a genotype may have in varying environments. No distinction is made between a species' natural environment and those environments containing interfering forces. The different phenotypes that may result from a genotype are just the upshot of that genotype occurring in different environments. Each phenotype is considered the natural result of its genotype in a particular environment.

Perhaps the dispositional account of species essentialism should be viewed as asserting that each species has a unique natural genotype. On this reading, organisms often lack their species' natural genotype because of such processes as mutation, recombination, meiotic drive, and random drift. Contemporary genetics, however, does not enshrine any particular genotype as the natural one for a species. Each genotype is considered the natural upshot of its parents' contributions. If a process like recombination occurs, the genetic results are considered natural as well. Indeed, the variability caused by recombination is generally considered the function, not the dysfunction, of sexual reproduction.

The above discussion shows that the dispositional account of species essentialism lacks a basis in contemporary biology. Essentialism has lost its theoretical basis in other ways. Recall that a primary motivation for essentialism is to explain the behavior of individuals within a kind. We have seen that the essentialist explanation of an organism's traits is inconsistent with population genetics. A similar problem occurs when we move one level up and consider the essentialist explanation of population-level properties.

Mayr (1959) illustrates this point by drawing the distinction between typological and populational thinking. A typologist is an essentialist, and a populationist is an evolutionist.

> The ultimate conclusions of the population thinker and of the typologist are precisely the opposite. For the typologist, the type (*eidos*) is real and the variation an illusion, while for the populationist the type (average) is an abstraction and only the variation is real. No two ways of looking at nature could be more different (Mayr 1959 [quoted in Mayr 1963, 5]).

The typologist explains the properties of a population by citing the natural tendencies of its individual members and the forces that interfere with those tendencies. For example, the typologist might explain why the majority of organisms in a population are tall by citing tallness as the natural tendency of that population's organisms and a lack of interference with that tendency. What is explanatorily primary for the typologist is the organisms' natural tendency or type.

With the advent of evolutionary theory, this kind of explanation of population phenomena is no longer needed. Instead of explaining population phenomena by citing essences and interfering forces, the populationist cites the frequency of traits in the previous generation of a population as well as the evolutionary processes that affect that population. So, for example, the populationist explains the frequency of tall organisms in one generation by citing the frequency of tallness in the previous generation and the occurrence of such processes as selection and mutation. The populationist does not, and need not, make any reference to the natural tendencies of organisms. Citing the gene frequencies of a population has replaced any reference to the natural states of organisms. Hence, population thinking has rendered the essentialist mechanism for explaining population phenomena theoretically superfluous.

Enough has been said to illustrate the death of species essentialism. We have seen that a variety of forces conspire against the occurrence of qualitative traits in all and only organisms of particular species. We have also seen that the mechanisms posited by essentialism to explain biological diversity are inconsistent with evolutionary theory. Essentialism is out of step with contemporary theory and is an inappropriate mode for classifying organisms into species. Notice that the arguments against essentialism here are empirically based. Our best empirical theories conspire against species essentialism. No a priori arguments concerning the problems of essentialism have been mustered. If essentialism should not be used for classifying organisms into biological taxa, then what general approach to classification should be used? Two general approaches remain to be considered: cluster analysis and the historical approach.

3.2   THE FAILURE OF CLUSTER ANALYSIS

As we saw in Chapter 1, cluster analysis requires that the members of a taxonomic unit share a cluster of similar traits. Yet unlike essentialism, cluster analysis does not require that a property occur in all and only the members of a taxonomic unit. Recall that one failure of essentialism is that rarely, if ever, do biological traits occur in all and only the members of a species. Cluster analysis does not place such a requirement on taxonomic units and thus avoids one of the pitfalls of essentialism. Should we therefore consider cluster analysis an appropriate mode for classifying organisms into species? To answer that question we need to examine the various forms of cluster analysis introduced in Section 1.2: the similarity approaches of Hempel and pheneticism; Wittgenstein's notion of family resemblance; and Boyd's account of homeostatic property cluster kinds.

All versions of the similarity approach require that the members of a taxonomic unit be more similar to one another than they are to entities in other units. An example of how one would apply the strong version of cluster analysis to species is offered by Dupré (1981, 82):

> The existence of species, I suggest, may be seen as consisting in the following fact. If it were possible to map individual organisms on a multidimensional quality space, we would find numerous clusters or bumps. In some parts of biology these clusters will be almost entirely discrete. In other areas there will be a continuum of individuals between the peaks. It can then be seen as the business of taxonomy to identify these peaks.

The first problem with this cluster approach is that it produces classifications that biologists, including pheneticists, would find implausible. Consider the different developmental stages of a single organism. A May beetle exhibits quite different morphological and behavioral traits when it is a larva, a pupa, and an imago (Hennig 1969, 6; the same is true of many insects and lower invertebrates; see Mayr 1963, 32). A cluster analysis based on overall similarity would sort the various developmental stages of May beetles into three different species. A single organism itself, in other words, would belong to different species at different stages of its life. No biologist intentionally assigns one temporal stage of an organism to one species and a later stage to another species. When a biologist does so and realizes what she has done, she quickly fixes her classification so that the organism belongs to a single species for its entire life. The genus *Procheneosaurus*, for example, was found to contain only the juvenile stages of organisms from two different species. When that was discovered, the positing of the genus *Procheneosaurus* was dropped (Wiley 1981, 33). Biologists attempt to sort entire organisms into species, not mere temporal stages of organisms. The cluster approach can be applied to species only if one alters the concept of species such that it has little relation to current use in biology.

It is not hard to find other cases of cluster analysis based on overall similarity wreaking havoc with common conceptions of species. In many groups of birds, insects, and lower invertebrates, the males of one species are morphologically more similar to the males of another species than they are to the females of their own species (Mayr 1963, 32). A classification based on overall similarity would, for example, place male hummingbirds and female hummingbirds into different species. Such a classification defies basic tenets of biology – it contains species of organisms that have no mode of reproduction, yet juveniles mysteriously appear. Again one could convert the term "species" so that it produces classifications according to overall similarity, even though some species consist of only males and others of only females. But the theoretical cost of such a move is not one biologists, even pheneticists (Hull 1988, 125), are willing to pay.

The above examples show the implausibility of sorting organisms into species according to overall phenotypic similarity. What about placing organisms into species according to overall genetic similarity? Matters get a bit more complicated at the genetic level. Local populations of the *Drosophila willistoni* complex tend to contain members with more overall genetic similarity than members in different reproductively isolated species (Futuyma 1986, 220); so, in that group of fruitflies, overall genetic similarity tends to go hand in hand with the ability to interbreed. However, as Futuyma (1986, 222) and Templeton (1981) note, sometimes reproductive barriers between organisms turn on genetic differences at just a few loci. In such cases, overall genetic similarity and the ability to reproduce successfully may identify different groups of organisms. Thus Mayr (1963, 544) writes:

> It is easy to imagine two conspecific populations that share the same species-specific isolating mechanism and essential chromosome structure, and yet differ from each other by more individual gene substitutions than some good species. It is evident that a purely quantitative approach may well be misleading. Nor can species differences be expressed in terms of the genetic bits of information, the nucleotide pairs of the DNA. This would be quite as absurd as trying to express the differences between the Bible and Dante's *Divina Commedia* in terms of the difference in the frequency of the letters of the alphabet used in the two works.[2]

If one wants to keep organisms that interbreed and produce fertile offspring in the same species, as even pheneticists desire, then overall genetic similarity can steer one the wrong way.

Stepping back, we can see that the application of cluster analysis based on overall similarity is undermined by the same factor that undermines species essentialism – the fortuitous nature of evolution. Recall that any trait found in all and only the organisms in one generation of a species can, for a variety of reasons, lose its universality and uniqueness in later generations. Biological evolution offers no safe haven for essentialism. The same goes for the cluster approach based on overall similarity. Evolution can cause the males of a species to be more similar to the males of another species than they are to the females of their own species. It can cause organisms that can successfully reproduce to have less genetic similarity than organisms that cannot successfully reproduce. In sum, both essentialism and cluster analysis based on overall similarity are confounded by evolution. Therefore, such qualitative approaches to classification are inappropriate for sorting organisms into species.

There are other forms of cluster analysis. One might try to apply Wittgenstein's notion of family resemblance to species. This is the tactic Beckner (1959, 24) adopts in his work. Recall the connection between classification and Wittgenstein's account of family resemblance (Section 1.2). Games, for instance, are classified according to a disjunction of traits, where each disjunct cites a number of traits. If capital letters represent traits, then a definition of "game" will look something like this: all games have (A and B and C) or (A and B and D) or (C and D and E), or (A and B and E), and so on. The actual definition of "game" will be more complicated because each disjunct will be a longer conjunction of traits and the disjunction itself will contain many more disjuncts. Indeed, the disjunction may be open-ended since new types of games may be devised in the future.

A definition of a species' name based on family resemblance would be a disjunction where each disjunct cites a combination of traits found among that species' members. We would describe *Drosophila arizonensis* as the species containing organisms that satisfy one of the following disjuncts: (A and B and C ...) or (A and C and D ...) or (C and D and E ...) or, and so on. Again, the disjunction is open-ended because future members of that species may evolve to have very different traits. An immediate benefit of the family resemblance approach is that it avoids some of the problems facing essentialism and an approach based on overall similarity. The family resemblance approach does not require the existence of a trait in all and only the members of a species; thus it escapes that pitfall of essentialism. It does not require that the members of a species be more similar to each other than they are to organisms in other species; so it does not, for instance, sort the different life stages of a single organism into different species.

Though the family resemblance approach escapes these problems, it has its own problems to contend with. First, cluster analysis based on family resemblance is a fairly empty approach to classification. The family resemblance approach does not tell us how to construct classifications, nor does it tell us which organisms belong to which species. It merely tells us how to describe the traits found among the members of a species *after* we have determined what those members are. In other words, classifications based on family resemblance merely describe previous taxonomic practice. The family resemblance approach provides no guidance in constructing classifications, let alone classifications of organisms and taxa.

A corollary problem is that the family resemblance approach lacks the theoretical virtues of other approaches to classification. Promoters of a cluster approach based on similarity (Hempel and the pheneticists) desire such classifications because they provide a means for making inferences (Section 1.2). If a group of entities can be sorted according to their similar properties, then various generalizations about those entities can be made. Such generalizations can then be used for prediction and explanation. Similarly, knowing the essence of a taxon enables one to predict and explain many of the other traits found among the members of that taxon. Turning to the historical approach, knowing an organism's membership in a taxon provides information about that organism's causal connections to previous organisms and taxa. That, in turn, allows one to explain some of that organism's traits. The family resemblance approach, however, lacks such virtues. It places no restraints on whether the members of a taxon should be qualitatively similar or causally connected. As a result, it offers no means for predicting or explaining the behavior of a taxonomic unit's members.

Nothing I have said forecloses the application of the family resemblance approach to biological classification. I have merely argued that nothing is gained by employing that approach. Therefore, it should be employed in biological taxonomy, or in any scientific taxonomy, only as a last measure.

A more promising version of cluster analysis is Boyd's homeostatic cluster approach (Section 1.2). The members of a homeostatic cluster kind have a cluster of properties that "characterize" the members of that kind (Boyd 1990, 374). Those properties are caused by the homeostatic mechanisms found in the members of a kind. The ability to dissolve in acid, for example, is one of the properties that characterizes the members of gold; and the subatomic structure of gold is the homeostatic mechanism responsible for gold's ability to dissolve in acid. Boyd's approach has the virtue of providing a causal account of the traits typically found among the members of a taxonomic unit. None of the other cluster approaches provide such an account. Furthermore, Boyd's approach does not fall prey to the pitfalls of essentialism. Membership in a homeostatic cluster kind does not turn on having a set of necessary and sufficient properties. In fact, the traits typically found among the members of a homeostatic cluster kind can vary; even the types of homeostatic mechanisms found among the members of a kind can vary (Boyd 1990, 374). Boyd's homeostatic account is also less restrictive than cluster approaches that require overall similarity. The members of a homeostatic cluster kind need not be more similar to one another than they are to the members of another kind. Given these virtues, Boyd's homeostatic cluster analysis seems well suited to provide a philosophical account of species and other biological taxa.

Unfortunately, Boyd's account suffers from a serious drawback. The positing of causal homeostatic mechanisms provides a nice causal account of why the members of taxonomic units have their typical traits. But Boyd's cluster analysis fails to provide an account of what makes a homeostatic mechanism a homeostatic mechanism for a particular kind. Consider an application of Boyd's account to species. According to Boyd (1991, 142), the ability to interbreed successfully is a primary homeostatic mechanism for sexual species. That assertion is fairly uncontroversial. It is worth noting, however, that the homeostatic mechanisms of a sexual species can vary at a time and over time. Mayr (1963, 106), for example, writes that "the total isolation between two different species is normally due to a great multitude of different isolating forces." Not only does a single organism have multiple isolating mechanisms, but the different members of a species may have varying isolating mechanisms. Splitter (1988) extends this observation by applying it to species over time. He suggests that the isolating mechanisms of a species can evolve such that different isolating mechanisms are employed in different generations of a species. A sexual species' cluster of homeostatic mechanisms, then, is best described by an open-ended disjunction of such mechanisms.

Given the diversity of homeostatic mechanisms within sexual species, we are left with a couple of questions. What determines whether a homeostatic mechanism is appropriate for a particular species? What, in other words, determines the disjunction of homeostatic mechanisms for a species? That information must be obtained elsewhere, namely, from prior taxonomic knowledge of the members and homeostatic mechanisms of a taxon. We must already know which organisms belong to a species and which homeostatic mechanisms are associated with those organisms. Only after we have that information can the homeostatic approach be used to provide a description of a taxon's characteristics and its homeostatic mechanisms. When applied to species, Boyd's homeostatic approach fails to tell us why certain organisms with certain homeostatic mechanisms are members of a species. In this way, Boyd's homeostatic approach and Wittgenstein's family resemblance account suffer from a similar weakness: they fail to provide an adequate account of what makes the members of a species taxon members of that taxon.

One might respond that the problem I have highlighted is not so pressing. That a philosophical approach to classification that relies on information from other sources is not so bad. After all, philosophical accounts of particular scientific classifications should rely on empirical theories. But in the case of species, Boyd's account relies on a different philosophical approach to classification from his own. What binds a cluster of homeostatic mechanisms together such that they are homeostatic mechanisms of a particular species is genealogy. In his discussion of species as homeostatic cluster kinds, Boyd recognizes the importance of genealogy. He writes that a species is "a processlike historical entity" (1990, 374). What is being employed here is the historical approach to classification, not a qualitative approach such as Boyd's. The historical approach highlights the causal relations among the members of a taxonomic unit. On Boyd's cluster analysis, similarities among the members of a kind, especially their similar homeostatic mechanisms, come first. Casuality does enter the picture for Boyd's account, but it is causal mechanisms *within* the members of a kind; no causal relations *among* the members of a kind are posited by the homeostatic account. Only the historical account gets to the heart of the matter and provides an adequate account of what makes the homeostatic mechanisms of a species mechanisms of that species. Consequently, we should depart from qualitative approaches to classification such as Boyd's and turn to the historical approach itself.[3]

3.3 TAXA AS HISTORICAL ENTITIES

None of the qualitative approaches to classification properly address the concerns of biological taxonomy. The reason is fairly simple. They all neglect Darwin's insight that we can understand the organic world's diversity only by studying its evolution. Of course pre-Darwinian promoters of qualitative approaches cannot be blamed for neglecting the Darwinian revolution. Nevertheless, any post-Darwinian application of a qualitative approach to biological taxonomy is out of step with a well-accepted tenet of Darwinism, namely, that an important way of explaining the frequency of a trait in a taxon is to cite the evolution of that taxon with respect to that trait. As we shall see, such evolutionary explanations require that local populations, species, and higher taxa form genealogical entities. Only the historical approach classifies entities according to genealogical relations. Thus, only the historical approach can serve as a basis for evolutionary explanations. That, in brief, is the argument that the historical approach to classification is *the* proper one for biological taxonomy. What follows are the details. (Variants of this argument can be found in Mayr 1963; Hennig 1966; Ghiselin 1969, 1974; Hull 1976, 1978; and elsewhere.)

First, we need a more precise characterization of evolutionary explanations. Kitcher (1993, 25–6) suggests the following:

> [T]he *Origin* offers a class of problem-solving patterns, aimed at answering families of questions about organisms by describing the histories of those organisms. The complete histories will always take a particular form in that they trace the modifications of lineages in response to various factors.... A Darwinian history for a group G of organisms between t_1 and t_2 with respect to a family of properties F consists of a specification of the frequencies of the properties belonging to F in each generation between t_1 and t_2.

Suppose we want to know why a trait is prominent among the current members of a population. An evolutionary explanation provides a Darwinian history that traces the history of that trait. Among the factors that might be cited are these: where and when that trait first appeared, what caused its inception, what forces caused it to become prominent, and what caused it to be passed down through the generations of a population. Depending on when the inception of that trait occurred, an appropriate Darwinian history may extend to an earlier generation of that population or it may extend to an earlier generation of another population of that species. An appropriate Darwinian history may even trace a trait's history to a much earlier and perhaps extinct taxon. Thus the Darwinian histories required for evolutionary explanations extend through populations, sometimes through species, and sometimes through higher taxa. The next question is why Darwinian histories require that populations and taxa form historical entities.

Evolutionary biology explains the occurrence of a trait in a taxon in two ways. The first type of explanation cites the processes responsible for the initial occurrence of traits – for example, mutation or recombination. The second type of explanation explains why traits become common or rare within a taxon; here the processes may be natural selection or random drift. The latter type of explanation involves the citation of a Darwinian history. In particular, a Darwinian history explains the frequency of a trait within a taxon by highlighting the evolution of that taxon with respect to that trait. Consider the evolution of a population through natural selection. On a population genetics account of evolution, a population evolves when its gene frequencies change from one generation to the next. Natural selection is the cause of such evolution when three criteria are met: the organisms in the population vary in their traits; that variation causes those organisms to have differential fitness; and those traits are heritable – they can be inherited by future generations. The third criterion is crucial for establishing that populations are historical entities. Suppose only the first two criteria for selection are met. For example (from Sober 1984b, 351), suppose the mature males in a troop of monkeys drive the juvenile males to the troop's periphery where they are more vulnerable to predators. Having the trait of being a juvenile monkey (in this troop) decreases one's fitness. But because the trait of being a juvenile male is not heritable, the action of the mature males does not cause the next generation of monkeys to have fewer juvenile males. There is no selection against being a male juvenile. A population evolves with respect to a trait only if that trait is heritable.

Heredity is the process that connects the generations of a population. Minimally, a trait is heritable when faithfully transmitted from parent to offspring. According to current biology, that transmission is achieved when traits are encoded in genetic material and that material is passed to offspring. In sexual species, mother and father each contribute genetic material to an offspring and that contribution occurs only if each parent is appropriately connected to the offspring. The requirement that parent and offspring are causally connected holds for both natural and artificial reproduction. Suppose that future technology allows genetic traits to be read from a parent and then encoded in a synthetic gamete. Even in that situation, where parent and offspring do not have physical contact, parent and offspring must be connected through an appropriate causal sequence of events. Heredity requires that parent and offspring be causally connected, either directly or indirectly.

We can now give the general argument for the historical nature of populations. Natural selection can cause a population to evolve with respect to a trait only if that trait is heritable. Traits are heritable only if some organisms in those generations are appropriately causally connected. A population, therefore, can evolve by selection only if its generations are causally connected. The step to populations being historical entities is now a small one. Any entity whose parts must be causally connected is a historical entity (Section 1.3). Evolutionary explanations require the generations of a population to be causally connected. Hence, evolutionary explanations require populations to be historical entities.

This argument can be further generalized in two ways. First, only selection has been highlighted as the cause of evolution, but other forces can cause changes in gene frequencies as well. A trait may become predominant because of random drift. Alternatively, a trait might prevail, despite its being selectively neutral, because it occurs in those organisms that have escaped an environmental catastrophe. In such nonselective instances of evolution, the proper evolutionary history still requires that populations form historical entities. A trait will rise to prominence in such cases only if it is heritable. And heritability, as we have seen, requires that the appropriate generations be causally connected.

We can also generalize the above argument to include species and higher taxa. A local population consists of conspecific organisms that are present at a particular time and place.[4] The above argument obviously applies to species that consist of merely a local population. Yet many species consist of geographically disconnected populations. The above argument applies to them as well. An evolutionary explanation for the occurrence of a similar trait in the various populations of a species cites the transmission of that trait from a common ancestral population. In such cases, heritability requires that a species form a single genealogical lineage with the ancestral population at the lineage's trunk and the current populations at the tops of the lineage's branches. The same reasoning applies to an evolutionary explanation of similarities among various species of a higher taxon: those traits occurred in a common ancestral population and subsequently were passed down to that taxon's current species. Again, heritability requires that the species of a higher taxon be genealogically connected. In sum, any evolutionary explanation of the distribution of traits among the populations of a species, or the species of a higher taxon, requires that such taxa be historical entities.[5]

The above argument places a fairly modest ontological requirement on taxa, namely, that taxa are historical entities. It says nothing about what conditions suffice to make a group of organisms a population, a species, or a higher taxon. Nevertheless the requirement that taxa are historical entities is strong enough to show that the historical approach to classification is *the* appropriate one for biological taxonomy. None of the qualitative approaches – essentialism, overall similarity, family resemblance, and homeostatic cluster kinds – classify entities according to the causal relations among them. Only the historical approach is attuned to such relations.

The arguments in this and the previous two sections establish two points concerning which mode of classification is appropriate for biological taxonomy. First, the qualitative approaches to classification are ill equipped to handle the evolving and often heterogeneous nature of biological taxa. Second, only the historical approach properly captures the historical nature of biological taxa. Having argued that the historical approach is the correct one for biological taxonomy, I would like to turn to two issues closely associated with that conclusion. One concerns further speculation on the ontological nature of species – in particular, whether species are individuals. The other is the relationship between the historical nature of taxa and the status of evolutionary theory as a scientific theory.

3.4 SPECIES AS INDIVIDUALS

The claim that biological taxa are historical entities includes the claim that species are historical entities; but often this latter claim comes under a different guise – the assertion that species are individuals. Unfortunately the term "individual" is ambiguous. Sometimes the thesis that species are individuals amounts to nothing more than the claim that the organisms of a species must be genealogically connected. Other times it means something more – for example, that species form cohesive wholes or that the parts of a species causally interact. In this section we take a further plunge into the ontology of biological taxa by examining the "species are individuals" thesis. As we shall see, whether species are individuals depends on what one means by "individual" and what groups of organisms one is willing to allow as species.

A minimal notion of individuality is found in Ghiselin (1987[1992], 363–4) and Rosenberg (1985a, 204). For these authors, an individual is simply a "spatiotemporally restricted entity." The parts of an individual cannot be scattered anywhere in time and space but must occupy a particular spatiotemporal region. Species certainly satisfy this requirement, because all taxa are genealogical entities. A species forms a genealogical entity when its various members are connected by appropriate hereditary relations. Such hereditary relations require that the organisms of a species form a continuous spatiotemporal entity: each organism of a species must come into contact with another member or its genetic material. This requirement places a restriction on the spatiotemporal location of a species' organisms. Given that species are genealogical entities, species are individuals on Ghiselin and Rosenberg's account of individuality.

Many authors contend that there is more to being an individual than being a spatiotemporally restricted entity. Hull, for example, writes:

> [I]ntegration by descent is only a necessary condition for individuality, it is not sufficient. If it were, all genes, all organisms and all species would form but a single individual. A certain cohesiveness is also required . . . (1976, 183; see also 1980, 313).

Similar suggestions are found in Wiley (1981) and Williams (1985). For example, Wiley writes that "individuals are restricted to particular spatiotemporal frameworks and have both cohesion and continuity" (1981, 74). For Hull, Wiley, and Williams, individuality requires both spatiotemporal continuity and cohesion.

Given the added condition of cohesiveness, a question immediately arises: What does it mean to say that an entity is "cohesive"? Unfortunately, the notion of cohesion is ambiguous. Nevertheless, two main uses occur in the literature. According to some authors, a species is a cohesive entity when its members act in similar ways to similar processes. For instance, Williams writes that a group of organisms is cohesive when its members react in a relatively similar manner to similar selection pressures (1970, 356–7). Likewise, Templeton (1989) suggests that a species is cohesive to the degree that its members will react in the same way to the introduction of a mutation (also see Mishler and Brandon 1987, 400).

Others offer a slightly different account of cohesiveness: they contend that a species is cohesive when a certain coherence or evolutionary unity is maintained among its members. For example, when Hull (1978[1992], 300) talks about the processes that cause cohesiveness, he asks, "Why do groups of relatively independent local populations continue to display fairly consistent, recognizable phenotypes?" Why are not asexual species "as diffuse as dust-storms in the desert?" Hull is not advocating a form of species essentialism. He is merely asserting that the organic world seems to consist of groups of organisms that share a common way of life – that the organic world seems to come in recognizable packages that we call "species." This notion of cohesion (sometimes referred to as "coherence" and "evolutionary unity") is used by numerous writers; see, for example, Wiley (1981, 74), Mayr (1970, 300), and Eldredge and Gould (1972, 114).

Perhaps these different meanings of cohesion are two sides of the same coin: the organisms of a species share a common way of life because their members tend to react in similar ways to similar processes. For my part, I find the usage of "cohesion," "coherence," and "evolutionary unity" in the literature too ambiguous (see Ereshefsky 1991a). My tack here will not be to provide an analysis of the notion of cohesion but to focus on the processes that may cause such cohesiveness.

Those who argue that species are individuals because they are cohesive entities cite three processes as promoting cohesion: gene flow, genetic homeostasis, and exposure to similar selection regimes. As we saw in Section 2.5, gene flow promotes a species' cohesiveness in two ways. Traits that arise within a population and prove to be adaptive are transmitted by gene flow to the other populations of a species. Gene flow thus prevents the divergence of a local population by spreading adaptive genes (Mayr 1970, 298, 300). Gene flow also prevents divergence by more subtle means, namely, by damping the occurrence of change in local populations (Mayr 1970, 168–9). Suppose a new gene arises within a population. That gene will become predominant only if, as Mayr puts it, it is able to "co-adapt" with the other genetic material in that population. The more genetic variability in a population, the more types of genetic material a new gene must be able to mix with. Because gene flow increases genetic variability within a population, it causes a new gene to be more stringently tested. So gene flow not only spreads traits from population to population but it also hinders the rise of new traits in local populations.

A second process that may cause species to be cohesive entities is genetic homeostasis. According to Eldredge and Gould, we should view species "as homeostatic systems – as amazingly well buffered to resist change and maintain stability in the face of disturbing influences" (1972, 114). The notion of homeostatic systems first comes from Waddington (1957) and is elaborated by Mayr (1970, 182, 300) as follows. Within a species there is a constant influx of new genetic material. This influx can destabilize the well-adapted phenotypes of a species. Given the influx of potentially destabilizing genetic material, selection occurs for genotypes that produce favored phenotypes despite the reconstruction of a species' genotypes. Furthermore, there is selection for genotypes that produce phenotypes that do well in changing environments. Genotypes that continue to produce well-adapted phenotypes, despite the reconstruction of genotypes and variation in the environment, are homeostatic. Due to their homeostatic nature, such genotypes work to preserve a species' unity.

Selection is a third process that promotes cohesion within species. Ehrlich and Raven (1969), among others (see Section 2.5), suggest that the members of a species are exposed to common selection forces. When powerful enough, these forces are responsible for a species' unity. There are two ways selection promotes cohesiveness in species. First, strong stabilizing selection can weed out new traits that might cause the members of a population to diverge. Second, selection can cause the prominence of advantageous traits within a species.

Hull (1976, 1978), Wiley (1981), and Williams (1985) suggest that these three processes – gene flow, genetic homeostasis, and selection – are important in making species individuals. For them, species are individuals because two conditions are met. First, species consist of genealogically connected organisms; hence, species form spatiotemporally continuous entities. Second, species are exposed to one or more of the processes cited above, and as a result, form cohesive entities.

A third notion of individuality lurks in the literature. According to Mishler and Donoghue (1982), Mayr (1987), Ruse (1987), Guyot (1987), Ereshefsky (1991a), and Sober (1993), there is more to being an individual than spatiotemporal continuity and cohesiveness. For these authors, the parts of an individual need to be spatiotemporally connected and *causally integrated*. Accordingly, a species forms an individual only if its organisms are spatiotemporally connected to a common ancestor and there is some appropriate causal interaction among its organisms and populations. Before turning to the question of whether such interaction occurs within species, let us examine the motivation for this notion of individuality.

On this account of individuality, an entity is an individual only if its parts are appropriately causally connected. As we saw in Section 1.3, not any causal connection will do. The parts of an individual need to be connected by those processes that make them parts of a particular type of individual. What determines which casual connections are appropriate and how frequently such interactions must occur? To answer these questions, we need to turn to the theories that study the entities in question (assuming such theories exist; see Section 1.3). This notion of individuality accords with intuitions concerning paradigmatic individuals and nonindividuals. Ghiselin (1969, 1974) and Hull (1976, 1978) suggest that organisms, countries, and planets are obvious examples of individuals. We tend to think that these entities are the types of entities they are because their parts must appropriately interact. Nonindividuals, such as the natural kind gold or all the bullets shot in World War II, consist of members that need not causally interact. If they do causally interact, such interaction is not necessary for their being members of those classes. So what seems to distinguish paradigmatic individuals and nonindividuals is that the parts of an individual must appropriately interact, whereas the members of a nonindividual need not causally interact.

Given this view of individuality, are species individuals? The answer to that question depends on what forces cause organisms to be the parts of species and what types of organisms form species. Species consisting of populations that are bound by gene flow satisfy the strong notion of individuality. But what percentage of species consist of populations that are held together by gene flow? Notice that there are two questions here. How many species consist of populations that exchange genetic material through gene flow? When gene flow is present, how effective is it in maintaining a species' unity? According to some biologists (for example, Ehrlich and Raven 1969; Eldredge and Gould 1972, 114; Lande 1980, 467; and Grant 1980, 167), gene flow is *not* the prominent force in maintaining the unity of species. These biologists offer several reasons.

First, species consisting of asexual organisms lack the process of gene flow. Genetic material is passed down through parent-offspring relations, but none is exchanged through intragenerational connections. Accordingly, gene flow is not a process that causes asexual species to be cohesive entities. Notice that the occurrence of asexual reproduction is not a biological oddity but the predominant form of reproduction in life on this planet. For the first three-quarters of life on Earth, asexuality was the only form of reproduction (Hull 1988, 429). Furthermore, if one looks at current biota, most organisms reproduce asexually; most plants and insects, not to mention fungi and microorganisms, reproduce asexually.[6]

Turning to organisms that reproduce sexually, debate exists whether gene flow binds those organisms into distinct species. Some biologists (Ehrlich and Raven 1969, Jackson and Pound 1979, Levin and Kerster 1974) maintain that many species of sexual organisms consist of populations that exchange either no genetic material or only a negligible amount. These authors, as well as others (Endler 1973; Lande 1980, 467; Grant 1980, 167), suggest that even where gene flow is extensive, gene flow may not be an effective stabilizing force in species. Their suggestion is based on empirical studies in the laboratory and in the field, which indicate that gene flow is neither necessary nor sufficient for maintaining a species' coherence.

Such considerations imply that the unity of many, perhaps most, species is the result of genetic homeostasis or exposure to common selective regimes rather than gene flow. Genetic homeostasis and selection are processes that do not require any interaction among the organisms or populations of a species. Individual selection acts independently on each organism, and genetic homeostasis works independently within each organism. So, if most species are cohesive entities as the result of such processes, and if one adopts the notion that individuals are causally interactive entities, then most species are not individuals.

Some supporters of the "species are individuals" thesis have responded to this suggestion. Eldredge (1985) and Ghiselin (1987, 1989), for example, argue that the existence of asexual organisms does not pose a threat to the individuality thesis because asexual organisms do not form species. Only populations of interbreeding organisms can form species. Ghiselin (1987[1992], 374) offers an analogy. Some workers pool their resources and form trade unions and some workers do not. Similarly, some organisms pool their genetic resources through interbreeding while some organisms (asexual ones) do not. Eldredge (1985, 200–1) and Ghiselin (1989, 74) highlight an important difference between sexual and asexual species. Recombination occurs only in sexual organisms and makes sexual species more genetically flexible than groups of asexual organisms. That increased flexibility helps sexual species better react to changing environments than groups of asexual organisms.

Does Ghiselin and Eldredge's response save the "species are individuals" thesis? I am doubtful. It does not address the existence of species consisting of interbreeding organisms whose populations are not connected by gene flow. No one denies that such species exist. In addition, Ghiselin and Eldredge's response does not address the question of whether the presence of gene flow is effective in unifying sexual species. Even Mayr (1970, 300), father of the most prominent inter-breeding-based species concept, writes that for sexual species, "[i]t is evident that gene flow alone is not enough to overcome entirely the local effects of mutation and selection." So even if one denies that asexual organisms form species, the question remains whether gene flow is a primary force in sexual species.

What about the suggestion that asexual organisms do not form species? I have two reservations with that suggestion. First, to deny that asexual organisms form species is to deny that most of life on this planet forms species. This seems too high a price to pay to retain the interbreeding approach to species and the individuality thesis. Second, and more important, it is far from clear how the difference in recombination suffices to establish that asexual organisms do not form species. As Mishler and Budd (1990) note, asexual and sexual organisms form equally stable lineages. Similarly, Templeton (1989[1992], 164) writes that "most parthenogenetic 'species' display the same patterns of phenotypic cohesion within and discontinuity between as do sexual species." In addition, the types of processes that underwrite the stability of sexual and asexual lineages, genetic homeostasis and selection, may be the same.

So are species individuals? It all depends on the notion of individuality adopted and the types of organisms allowed as members of species. For my part, I am sympathetic to the third and more restrictive notion of individuality – that the parts of an individual must appropriately interact. Furthermore, I am willing to allow that asexual organisms form species. Given these two predilections, I would suggest that most species are not individuals. Notice, however, that disagreements over the nature of individuals and whether asexual organisms form species do not affect the claim that species are historical entities. All three notions of individuality surveyed here concur that species are genealogical entities.

3.5 THEORY AND EXPLANATION

The historical nature of taxa constrains the proper way to construct biological classifications. An approach to classification emphasizing causal connections, rather than one highlighting qualitative similarities, is the appropriate one for biological taxonomy. The historical nature of taxa is thought to have further implications. One implication concerns the scientific status of evolutionary theory. In brief, the argument for that implication goes something like this: If species are historical entities, then there are no natural laws about particular species. If that is the case, then evolutionary biology cannot provide law-based explanations similar to those found in physics and chemistry. As a result, evolutionary biology is a different sort of discipline from physics or chemistry. In this section, several versions of this argument are reviewed. I argue that the historical nature of taxa does not cause explanations in evolutionary biology to differ from those in the physical sciences.

The best place to start is with the standard philosophical account of scientific explanation: Hempel's (1965) covering-law model of explanation. According to that model, a good scientific explanation is special type of logical argument. Its premises cite the relevant initial conditions and scientific laws, and its conclusion describes the event being explained. An explanation for the speed of a falling object, for instance, cites the height from which that object is dropped, the current wind speed, and other initial conditions, as well as Newton's law of gravitation and other relevant laws. Together these initial conditions and laws imply the object's speed and thereby explain why the object travels at the speed it does. This is an example of a nonstatistical or deterministic explanation. The covering-law model allows for statistical explanations as well. Such explanations cite those statistical laws and initial conditions that confer a high probability on the event being explained.

The covering-law model of explanation is not without its problems (see Salmon 1989 for a survey). In response to those problems a number of alternative accounts of explanation have been offered (again, see Salmon 1989). Although these alternatives depart from Hempel's, most require that explanations cite scientific laws. Consider the most widely accepted alternative, Salmon's (1984) causal model of explanation. The causal model varies from the covering law model in several ways. Most important, it requires that the initial conditions cited in an explanation *cause* the event being explained. Still, the causal account requires that adequate explanations cite laws of nature – in particular, causal laws.

Much more can be said about philosophical accounts of scientific explanation (in fact, many books have been written on the topic). Our interest here is merely the requirement that scientific explanations contain laws. It is that requirement that bears on whether evolutionary theory is a different sort of theory from those in physics and chemistry. But before seeing whether the historical nature of biological taxa causes evolutionary theory to lack laws, we need to say a little more about the nature of laws.

Again, the philosophical literature is vast and contains numerous proposals. Nevertheless, philosophers often place four requirements on the nature of scientific laws. First, laws are characterized by either universal generalizations ("All A are B") or statistical generalizations ("All A have probability X of being B," where $0 < X < 1$). Second, such generalizations cannot be true merely in virtue of their syntax or the meanings of their words. Instead, a law of nature must be empirical. Neither "All apples are apples" nor "All bachelors are unmarried males" are laws of nature. Third, laws cannot make an uneliminable reference to a particular time and place. They should have a spatiotemporally unrestricted scope. "All the paper currency in my wallet on May 3, 1994, is green, white, and black" is not a scientific law. On the other hand, Galileo's law of the pendulum ("the period of a pendulum is proportional to the square root of its length") makes no reference to a particular time or place. Fourth, laws are *counterfactually* true. That is, they are true even if there are no instances of the entities or processes described by them. "All copper conducts electricity" is true even if, contrary to fact, there is no copper in the universe. These four characteristics of laws are not without controversy, nor do they exhaust the list of characteristics suggested by philosophers. (See van Fraassen 1989 for a survey of what philosophers have to say about scientific laws.) Nonetheless, these characteristics capture the notion of law presupposed by the covering-law model of explanation.

Let us turn to the question of whether evolutionary biology lacks scientific laws. Two arguments are generally offered to support the conclusion that evolutionary theory lacks laws; both arguments stem from the historical nature of taxa. The first argument cites the evolutionary nature of taxa, namely, that taxa are vulnerable to a number of evolutionary forces that block the universality of a biological trait within a particular taxon. As we saw in Section 3.1, rarely are biological traits found in all the members of a particular taxon. When they are, their universality can easily be destroyed in subsequent generations due to mutation, recombination, random drift, and environmental change. Evolutionary theory enshrines no biological trait as necessary for membership in a taxon. Instead, biological taxa are heterogeneous, evolving lineages of organisms in which every genetically based trait is vulnerable. Evolutionary theory, therefore, offers no universal generalizations about the members of particular taxa akin to those concerning the chemical elements in chemistry. (Versions of this argument can be found in Smart 1963, 54; 1968, 93ff.; and Beatty 1980, 549–50; 1981, 407.)

One might respond that though there are no universal laws concerning the members of particular species, still there might be statistical laws. Recall that the covering-law model of explanation allows that the laws cited in explanations may be universal *or* statistical. This is just the tactic Dupré (1993, 82) and Matthen (1998, 118) employ when they argue that evolution does not prevent the existence of laws concerning particular taxa. However, the shift to statistical laws does not help. Laws, statistical or otherwise, are supposed to be stable: a generalization is a law of nature only if it is true in the past, in the present, and in the future. The evolutionary nature of taxa spells trouble for any attempt to find stable statistical generalizations concerning particular taxa. Suppose after several years of observation we find that 70 percent of toucans start their mating ritual in August. The various evolutionary forces cited in the previous paragraph can alter that percentage in future years. Maybe 70 percent of today's toucans start their mating rituals in August, but in five years it may be 30 percent, and after that 50 percent. Even statistical generalizations about particular taxa are vulnerable to the whims of nature.[7]

The second argument for evolutionary biology lacking laws concerns the contingent nature of taxa. Evolutionary events that affect entire taxa – for example, the extinction of the dodos or the origination of vertebrates – depend on complex suites of factors. The probability of such combinations of factors occurring again is extremely low. Because such events are not likely to be repeated, they should be considered singular or unique. But if they are singular, then no laws apply to them because a law must range over more than one event to be a genuine law of nature. According to the authors that press this line of reasoning (Goudge 1961, 65ff.; Simpson 1964, 186; Mayr 1982, 71–2; Gould 1986), such evolutionary events are not explained by laws but by citing the sequence of events that precede them. Explanations in evolutionary biology, they suggest, do not fit the covering-law model but fall under the "historical narrative," "particular-circumstance," or "integrating" models of explanation. (A more thorough discussion of these alternative models of explanation is provided later in the chapter.)

Philosophers and biologists draw two conclusions from these arguments. For Smart (1963, 1968), the absence of laws in evolutionary theory shows that evolutionary biology is not a genuine scientific enterprise. For others (Goudge 1961, Gould 1986, O'Hara 1988, Mayr 1988), the absence of laws does not show that evolutionary biology is either inferior or unscientific; it merely shows that evolutionary biology is methodologically different. Both conclusions assume that evolutionary theory is a different sort of theory from those in physics or chemistry. In what follows, I will challenge this assumption. In particular, I will suggest that the historical nature of biological taxa does not force biologists to employ a distinctive mode of explanation.

According to the first argument, taxa are exposed to a number of evolutionary forces. Those forces prevent the existence of universal or stable statistical generalizations concerning particular taxa. Thus evolutionary theory lacks law-based explanations. Let us assume that no laws about particular species exist. Does that show that there are no laws in evolutionary biology? At first glance it might, if one assumes that species are qualitative kinds akin to the elements of the periodic table. But as Hull (1977, 1978, 1987) has argued, if taxa are historical entities, then the absence of laws ranging over the members of a particular taxon does not show that there are no laws in evolutionary biology. To think that it does is to confuse the historical entities (or particulars) of a theory with its qualitative kinds. Consider an analogous situation in chemistry and geology. Chemical and geological laws range over the members of qualitative kinds, not over the parts of particular historical entities. There are no laws in those theories concerning all the parts of a particular rock. Yet one does not charge that those theories lack laws. The laws of those theories lie at different ontological levels; they concern kinds of rocks (granite or quartz) and kinds of elements that compose those rocks (iron or silver).

Similar considerations apply to evolutionary biology. There are no laws concerning all the members of particular species, but that does not imply that evolutionary biology lacks laws. Species and biological taxa are historical entities. If there are laws in evolutionary biology, they exist at other ontological levels. Perhaps laws range over kinds of species, kinds of populations, or kinds of organisms, where such groups are based on qualitative similarities rather than genealogical relations. Hull (1987, 173–4) and Ghiselin (1989, 58) suggest that the class of peripheral isolates (founder populations) forms a theoretically important kind in evolutionary biology. Hull (1987, 174) also suggests that some classes of species – for example, the class of polytypic species and the class of cosmopolitan species – form qualitative kinds. Further candidates of qualitative kinds in evolutionary biology are offered in Sober (1984a), Ghiselin (1989), and Ereshefsky (1995b). The important point here is that the absence of laws concerning particular taxa does not by itself imply the nonexistence of laws in evolutionary biology. One must also examine whether laws range over the qualitative kinds of evolutionary biology.

One might concede that there are generalizations that range over some qualitative kinds in evolutionary biology but still argue that those generalizations have exceptions and thus are different from the exceptionless laws found in the physical sciences. Consider Ghiselin's (1987, 1989) suggestion that Mayr's hypothesis that speciation usually starts when a small population is isolated from the main body of a species constitutes a law. "Mayr's law," as Ghiselin calls it, goes something like this: "speciation does not occur without an initial period of extrinsic isolation, such as a geographic barrier" (Ghiselin 1987, 129). There is empirical evidence that speciation tends to start only when a small population is isolated from the main body of its species (Carson and Templeton 1984; Futuyma 1986, 240–2). But as Mayr (1970, 254–6) recognizes, not all speciation starts that way. Some speciation events are the result of polyploidy (the multiplication of normal chromosome number in meiosis), and such speciation does not require the existence of isolated founder populations. So Mayr's law has exceptions.

Does that imply that Mayr's law is a different sort of law from those found in the physical sciences? Scriven (1961), Cartwright (1983), and Giere (1988) have argued that the laws of physics are also riddled with exceptions. Consider Galileo's law of the pendulum: the period of the pendulum is proportional to the square root of its length. Many real pendulums fail to conform to this law due to such interfering factors as air resistance, surface friction, and nonuniform gravitational attraction. The same applies for Galileo's law of free fall and numerous other physical laws. Both evolutionary laws and physical laws have exceptions. So the existence of laws with exceptions in evolutionary biology does not mark a difference between evolutionary biology and the physical sciences.

It should come as no surprise that the laws of physics and biology contain exceptions. The laws highlighted in science textbooks tend to be simplified descriptions of putative regularities in the empirical world. Such laws cite what are considered to be the important causal factors at work, for example, gravity in pendulums and geographic isolation in speciation. For convenience sake, laws tend not to mention the various background conditions that must occur if those laws are to be accurate. For instance, Galileo's law of the pendulum does not cite the interfering forces that may cause a pendulum to depart from a quantity predicted by the law. Similarly, Mayr's law does not cite other factors besides geographic isolation that may give rise to speciation. Physical and biological laws when taken literally and without attention to their implicit or explicit ceteris paribus provisos certainly have exceptions.

I have tried to close the gap between evolutionary biology and the physical sciences by suggesting two things. First, evolutionary biology may contain qualitative kinds which serve as grounds for lawlike generalizations. Second, the laws in both the physical sciences and evolutionary biology have exceptions. These points address the argument that the evolutionary nature of taxa prevents the existence of laws in evolutionary biology akin to those in the physical sciences. Still, the second argument for the difference between evolutionary biology and the physical sciences remains. Recall the gist of that argument. Evolutionary events concerning entire taxa are often caused by numerous factors. Such combinations of factors will never likely happen again; thus, such evolutionary events are unique. If they are unique, then no laws can be cited to explain them, for a law must range over more than one event to be a genuine law of nature. Instead, such unique evolutionary events are explained by citing the sequence of events leading up to them rather than any law of nature. The difference, then, between the physical sciences and evolutionary biology is that the former provide law-based explanations while the latter provides explanations that cite no laws.

There are several ways to respond to this second argument. One is to maintain that laws of nature are still laws even if they have only one instance. Indeed, they may lack any instances. Recall that laws are counterfactually true. Their generalizations are conditional statements that specify what would happen *if* certain conditions were met. The truth of such conditional statements does not turn on their antecedents being satisfied (Ereshefsky 1991b). For example, Galileo's law of the pendulum remains true even if no pendulums exist. Another line of reasoning questions the claim that complex evolutionary events cannot be captured by theoretically significant generalizations (Ruse 1973, 62ff.; Hull 1974, 98–9). Perhaps such complex events can be captured by highlighting just some of their important features. For instance, it might be the case that the extinction of dodos can be captured by a generalization that merely cites what happens to a species when it cannot escape from its predators. A third response, the one I will pursue, takes a different tack. For argument's sake, let us grant that a number of explanations in evolutionary biology fail to cite any laws and there are no plausible candidates for laws that might underwrite such explanations. Would the existence of such explanations set evolutionary biology apart from physics?

Before answering that question, we need a clearer picture of explanations that do not rely on laws. Goudge (1961), Hull (1992), and Richards (1992) call such explanations "historical narratives," "integrating explanations," and "particular-circumstance explanations." Such explanations explain a target event by citing the relevant events leading up to it and the causal connections that bind those events. An example of such an explanation is Gould's (1980) account of why contemporary pandas have false thumbs fashioned from their sesamoid bones. Gould cites the relevant events in the genealogical history of today's pandas. Ancestral carnivorous pandas lost their true anatomical first digit due to its limited use in clawing. Descendant herbivorous pandas subsequently developed a sesamoid "thumb" for stripping leaves off bamboo shoots. That trait proved to be adaptive, was selected for, and eventually became prominent among pandas. This explanation is a "historical narrative" because it provides a narrative of the relevant part of the panda's history. It is an "integrating explanation" because it integrates (binds) a series of events into one ongoing historical event – the evolution of pandas. Furthermore, this explanation is a "particular-circumstance explanation" because it cites only the particular events that led up to the event being explained.

Suppose evolutionary explanations make no reference to laws. Would the existence of such explanations indicate that evolutionary biology is a different sort of enterprise from physics? I doubt that it would because physicists also offer explanations that fail to cite laws. Consider the current explanation of background radiation in the universe (Faye 1994). In 1965, Arno Penzias and Robert Wilson detected a small amount of energy at a uniform temperature coming from every region of space. This phenomenon was subsequently explained as residual energy from the Big Bang. In brief, here is the explanation. The Big Bang created the universe's neutrons. Shortly after the Big Bang, neutrons started decaying into protons, electrons, and neutrinos. Some protons and neutrinos recombined to create neutrons and positrons. Some of those positrons collided with electrons and in doing so gave off the background radiation detected by Penzias and Wilson. In this explanation, only the events leading up to the existence of background radiation are cited. No physical laws exist that assert that when universes are created they are initially created with neutrons. Furthermore, no laws exist that imply that just this sequence of events will occur. Consequently, only a historical narrative can be cited to explain the existence of background radiation. (See Faye 1994 for other examples of explanations in the physical sciences that do not cite laws.) Physics, like evolutionary biology, contains explanations for which particular circumstances rather than laws carry the explanatory weight.

In this section I have tried to blunt the charge that the historical nature of taxa renders evolutionary biology a different type of discipline from those in the physical sciences. As we have seen, the lack of stable generalizations concerning all the members of a particular taxon does not indicate that evolutionary biology is void of the sorts of laws found in physics. Both disciplines contain exception-ridden generalizations concerning qualitative kinds. Furthermore, both disciplines employ particular-circumstance explanations to explain complex phenomena. Of course, nothing I have said rules out there being any methodological difference between evolutionary biology and the physical sciences. I have merely argued that the genealogical nature of species and higher taxa does not cause a methodological divide between those disciplines.

II

The Multiplicity of Nature

4

Species Pluralism

We have seen that the essentialist approach to classification does not apply to biological taxa (Section 3.1). Membership in the species *Drosphila melanogaster*, for instance, does not require that all and only the members of that species share a particular set of qualitative properties. Instead, the organisms of that species must be appropriately causally connected. Species *taxa* lack essences. But what about the species *category*? (Recall that species taxa are individual taxa, such as *Homo sapiens* and *Drosphila melanogaster*; the species category is the class of all species taxa.) Is there some property that all and only species taxa must have to be members of the category "species"? Is there, in other words, an essence to the species category?

Chapter 2 introduced no less than seven prominent species concepts. Each concept highlights a different set of biological properties that a taxon must have to be a species. Each concept, in other words, suggests a different essence of the species category. What is one to make of the diversity of species concepts in the literature? Biologists and philosophers have responded in two ways. Monists consider the species problem an unfinished debate in which the improper definitions need to be weeded from the proper one. Pluralists, on the other hand, maintain that no single correct definition of the species category exists. They suggest that we accept a number of species concepts as legitimate.

This chapter takes up the cause of species pluralism. Many authors maintain that evolution renders essentialism concerning species taxa outdated. Far fewer are willing to allow that evolution renders essentialism concerning the species category obsolete. The aim of this chapter is to motivate that additional step. The argument for species pluralism offered here comes in several stages. Section 4.1 motivates the initial plausibility of species pluralism, whereas Section 4.2 provides an ontological argument for its acceptance. Other authors have championed species pluralism and offer their own versions of pluralism (Ruse 1969, 1987; Dupre 1981, 1993; Mishler and Donoghue 1982; Kitcher 1984a, 1984b, 1987; Mishler and Brandon 1987; Stanford 1995). Section 4.3 surveys those forms of species pluralism and illustrates why the version offered here is preferable. As might be expected, monists are not happy with the promotion of species pluralism. They have responded with a series of objections to pluralism (Ghiselin 1969, 1987; Sober 1984a; Mayr 1987; Hull 1987, 1989). Section 4.4 answers those objections and poses an important methodological challenge to pluralism.

4.1 THE POSSIBILITY OF PLURALISM

Many of the species concepts surveyed in Chapter 2 fall into three general approaches to species: interbreeding, phylogenetic, and ecological. Not all of the species concepts reviewed in Chapter 2 fall under these three approaches. The phenetic species concept and various pattern cladist concepts do not. (A discussion of these concepts occurs in Section 5.5.) This section will not provide an exhaustive survey of various approaches to species. Instead, it will demonstrate that several prominent approaches to species provide very different pictures of the organic world. With that information in hand, we can then turn to the ontological argument for pluralism in the next section.

A number of species concepts fall under the interbreeding approach. Two prominent ones are Mayr's (1970) biological species concept and Paterson's (1985) mate recognition concept. According to these concepts, species are groups of organisms that interbreed and produce fertile offspring. A species' stability is due to interbreeding among its members. Of course, Mayr and Paterson have their differences. For Mayr, reproductive isolating mechanisms prevent the organisms of different species from successfully interbreeding. For Paterson, mate recognition systems are responsible for successful conspecific mating. Furthermore, Mayr proposes that reproductive isolating mechanisms are accidental by-products of selection for newly adapted genotypes whereas Paterson suggests that mate recognition systems are the result of selection for mechanisms that allow similarly adapted organisms to recognize each other as potential mates. Despite these differences, Mayr's and Paterson's concepts capture the heart of the interbreeding approach: a species taxon is a group of biparental organisms that share a common fertilization system. Other species concepts that fall under the interbreeding approach include those proposed by Dobzhansky (1970), Ghiselin (1974), Carson (1957), and Eldredge (1985).

The ecological approach provides a different picture of species. Advocates of that approach (Ehrlich and Raven 1969, Van Valen 1976 [1992], Andersson 1990) believe that environmental rather than reproductive factors are primarily responsible for the stability of species. They cite numerous examples that, they argue, illustrate the primacy of selection over interbreeding in maintaining species. One set of examples highlights species consisting of geographically isolated populations that do not exchange genetic material. Those species nevertheless remain ecologically, genetically, and morphologically distinct (see Section 2.5 for examples). Given the existence of such species, this question arises: If gene flow does not bind their populations, then what mechanism is responsible for the stability of such species? The proponents of the ecological approach suggest that the members of geographically dispersed species are exposed to similar selection forces that maintain the unity of such species. Another type of example cited in favor of the ecological approach highlights species taxa whose members frequently hybridize. A standard example is a multispecies (or "syngameon") consisting of distinct species taxa (see Section 2.5 and this chapter). The organisms of a multispecies often interbreed and produce fertile offspring, yet its component species are maintained as distinct evolutionary units because of species-specific stabilizing selection. For proponents of the ecological approach, the taxa that make up a multispecies are species, not the multispecies itself.

Although a handful of biologists promote an ecological approach to species, only one provides an ecological species concept. Van Valen (1976[1992], 70) offers the following definition of the species category: "A species is a lineage . . . which occupies an adaptive zone minimally different from that of any other lineage in its range and evolves separately from all lineages outside its range." The set of selection forces within a species' adaptive zone maintains a species as a distinct taxon, even if there is frequent hybridization between that species and another species within its range. It is worth noting that on the ecological approach species must be lineages and not merely groups of organisms that occupy the same adaptive zone. Species taxa are historical entities maintained and segmented by ecological forces.

The third general approach to species, the phylogenetic approach, is found in the work of process cladists. As we saw in Section 2.3, Hennig believes that descent from a common ancestor – the bare bones of genealogical evolution – is the primary process responsible for the existence of taxa. Accordingly, Hennig argues that classifications should highlight only those groups that are the result of common descent. Taxa should be "monophyletic." Recall that a monophyletic taxon contains an ancestral species and all of its descendant taxa. This description of monophyly highlights not only the process of common descent; it also highlights that taxa are "complete systems of common ancestry" (de Queiroz and Donoghue 1988, 318). Monophyletic taxa are complete systems of common ancestry whereas paraphyletic and polyphyletic taxa are not (see Section 2.1 for a discussion of these types of taxa).

Hennig intended his description of monophyletic taxa to apply to higher taxa and not species. Proponents of the phylogenetic approach want to extend Hennig's insight concerning the importance of common descent to species (Mishler and Donoghue 1982, Mishler and Brandon 1987, Ridley 1989). Thus they require that species taxa be monophyletic. Advocates of the phylogenetic approach agree that species are monophyletic, but they disagree on how to distinguish species taxa from all other types of monophyletic taxa. In other words, they disagree on a ranking criterion for species taxa. According to Mishler and Brandon (1987, 46), a taxon is "ranked as a species because it is the smallest 'important' lineage deemed worthy of formal recognition, where 'important' refers to the action of those processes that are dominant in producing and maintaining lineages in a particular case." For Ridley (1989, 3), a taxon is ranked as a species when it is a "set of organisms between two speciation events, or between one speciation event and one extinction event, or that are descended from a speciation event." Although proponents of the phylogenetic approach disagree on ranking criteria, they agree that propinquity of descendant is the primary organizing process of the organic world.

The three general approaches to species just highlighted – the interbreeding, ecological, and phylogenetic – are diverse, yet all three assume that species are lineages. Roughly, a lineage is either a single descendant-ancestor sequence of organisms or a group of such sequences that share a common origin. In philosophical jargon, these approaches assume that species taxa are spatiotemporally continuous or historical entities. As a result, the interbreeding, ecological and phylogenetic approaches satisfy one of the conclusions of Chapter 3, namely, that any post-Darwinian account of taxa requires that all taxa, including species, form historical entities. What is meant by "lineage" and "historical entity" can be illustrated further by using the distinction between monophyletic, paraphyletic, and polyphyletic taxa cited in Chapter 2. The three general approaches to species, when taken as a group, require that species taxa form either monophyletic or paraphyletic taxa. (The phylogenetic approach requires that species form monophyletic taxa; the interbreeding and ecological approaches require that species form either monophyletic or paraphyletic taxa.) All three approaches prohibit species from being polyphyletic taxa. Recall that the members of a polyphyletic taxon do not form a continuous genealogical sequence of organisms, whereas the members of monophyletic and paraphyletic taxa do. (This is graphically illustrated by Figures 2.2, 2.3, and 2.4.)

Beyond the assumption that species are lineages, the three approaches offer quite different accounts of species. Furthermore, these approaches give rise to classifications that cross-classify the world's organisms. Consider a hypothetical example. Suppose we want to classify the flies living in a marsh. The insects form three populations, A, B, and C (Figure 4.1a). Each population forms a basal monophyletic taxon. The organisms in B and C share a common ecological niche; the organisms in A occupy their own niche. Turning to breeding behavior, the organisms in A and B can successfully interbreed and produce fertile offspring. But the organisms in C reproduce asexually; their females reproduce via parthenogenesis, so their eggs do not require fertilization. What, then, is the correct classification of flies in this marsh? According to the phylogenetic approach, the correct classification consists of three taxa: A, B, and C (Figure 4.1b). According to the ecological approach, it consists of two taxa: A and B + C (Figure 4.1c). According to the interbreeding approach, it is a classification consisting of a single species A + B (Figure 4.1d). The organisms in C do not form an interbreeding species because they reproduce asexually. Notice, also, that the interbreeding species A + B and the ecological species A cannot form phylogenetic species because they are paraphyletic taxa. In sum, these three approaches to species provide three different classifications of the flies in the marsh.

Consider this situation on a much larger scale: trying to provide a classification of all the organisms on this planet. Because biologists disagree on the correct approach to species, they provide different classifications of the organic world. Moreover, those approaches

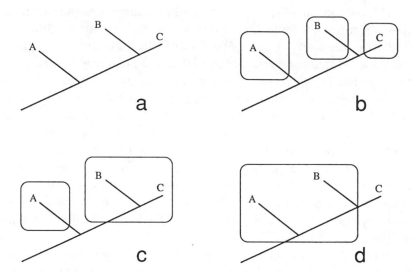

Figure 4.1   A phylogenetic tree with three populations: A, B, and C. (a) The phylogenetic tree. (b) The tree with three phylogenetic species: A, B, and C. (c) The tree with two ecological species: A and B + C. (d) The tree with one interbreeding species: A + B.

cross-classify the world's organisms by placing the same organisms in different species taxa. Cross-classification occurs in two ways (and can be illustrated using Figure 4.1). First, an organism may belong to two lineages with one lineage properly contained in another. For example, a member of the phylogenetic species A is also a member of the interbreeding species A + B. Second, an organism may belong to two lineages that are disjoint. For example, an organism in population B belongs to both the ecological species B + C and the interbreeding species A + B.

The type of pluralism highlighted here should be distinguished from a more modest form in the literature. Mishler and Donoghue (1982) and Mishler and Brandon (1987) require that all taxa be monophyletic, but they allow that various processes are responsible for the existence of species taxa. Some taxa are ranked as species as the result of interbreeding, others are the result of stabilizing selection, and still others the result of genetic homeostasis. Although they allow a plurality of ranking processes, Mishler, Donoghue, and Brandon maintain that an organism belongs to only one species. Consequently their pluralism provides a single classification of the organic world. The type of species pluralism offered in this chapter is more radical. It allows that varying approaches to species often classify the same organisms into different species taxa. The resulting picture is a multiplicity of classifications that cross-classify the organic world.

The marsh example given above is hypothetical, yet it nicely illustrates how various species approaches provide differing classifications. Still, one might wonder if those approaches provide varying classifications of the real world. The following examples demonstrate that they do. Let us start with the ecological and the interbreeding approaches to species. The biological literature cites a number of cases in which classifications based on these two approaches part company. Consider groups of organisms that Grant (1981) refers to as "syngameons" (some authors call them "multispecies," as shown earlier). A syngameon functions like a species on the interbreeding approach: its members interbreed and produce fertile offspring, and the members of different syngameons are reproductively isolated. On an ecological approach, however, syngameons consist of multiple ecologically distinct species. A classic example is the syngameon consisting of cottonwoods and balsam poplars. According to Templeton (1989[1992], 166), cottonwoods and balsam poplars have remained genetically, morphologically, and ecologically distinct for 12 million years despite their forming a syngameon. Other examples of syngameons occur in the genus *Quercus* (oaks) (Van Valen 1976[1992], 72), the genus *Iris* (Futuyma 1986, 115), the grass genera *Elymus*, *Stianion* and *Agrophyron* (Stebbins, Valencia, and Valencia, 1946), and many subfamilies of orchids (Holsinger 1984, 302).

The interbreeding and ecological approaches also give rise to varying classifications when asexual organisms form ecologically distinct lineages. The organisms of such lineages form species on the ecological approach but fail to do so on the interbreeding approach given their mode of reproduction. Examples of such lineages are found in the plant genera *Taraxacum*, *Hieracium*, and *Rubus* (Mishler 1990, 95). The organisms of these genera reproduce by apomixis: their females produce eggs that do not require fertilization. These genera nevertheless contain ecologically differentiated lineages of organisms recognized as species by botanists. This discrepancy between the interbreeding and ecological approaches is widespread since apomixis is common in higher plants, particularly in ferns and angiosperms (Briggs and Walters 1984, 127).

Just as the interbreeding and ecological approaches provide conflicting classifications, so do the interbreeding and phylogenetic approaches. The phylogenetic approach requires that all species taxa be monophyletic, and evidence of monophyly is provided only by shared derived characters (synapomorphies). Yet some interbreeding species form paraphyletic groups where the ability to interbreed successfully is a shared ancestral character (a symplesiomorphy) (Rosen 1979 and Donoghue 1985). In such cases, a taxon may be recognized as a species by the interbreeding approach but fail to be one on the phylogenetic approach. Donoghue (1985, 177) nicely describes such a situation. "Organisms that can interbreed are not necessarily closely related genealogically, but instead may be simply left-overs after some subgroup evolved a reproductive isolating mechanism." Examples of this phenomena occur in the freshwater fish groups *Xiphophorus* and *Heterandria* (Rosen 1979, 275–7). For instance, the members of C and F in *Xiphophorus* successfully interbreed and are reproductively isolated from organisms in other closely related groups (Figure 2.12). Thus the members of C and F form a species on the interbreeding approach, but because their ability to interbreed is a shared ancestral characteristic, the group C + F is paraphyletic and not considered a species on the phylogenetic approach.

Another example showing where the interbreeding and phylogenetic approaches part company concerns ancestral species. Suppose a handful of organisms from species A form a peripheral isolate and that isolate gives rise to a new species, B (Figure 4.2). A is considered an ancestral species because some of its members are founders of species B. On the interbreeding approach, ancestral species can survive the formation of offspring species as long as an ancestral species remains a distinct interbreeding unit. The phylogenetic approach does not allow the existence of ancestral species because such taxa are paraphyletic. Species A is a paraphyletic taxon because it contains some but not all of the descendants of A (it lacks the members of B). Here, then, is an important discrepancy between the interbreeding and phylogenetic approaches: an interbreeding unit that survives a speciation event remains the same species on the interbreeding approach, whereas on the phylogenetic approach it becomes a different species.

How often does this discrepancy occur in the organic world? It occurs whenever allopatric speciation via peripheral isolation occurs. As we saw in Section 2.5, some biologists believe that this type of speciation is the prominent form of speciation in nature. So the discrepancy between the interbreeding and the phylogenetic approaches may

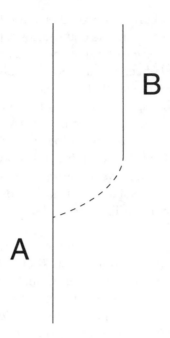

Figure 4.2   A geographically isolated population of species A becomes a new species, B. If species A continues to exist after the inception of species B, then A is a paraphyletic taxon.

be quite frequent. Another form of speciation in which an ancestral species may survive is speciation via polyploidy. Polyploids have a different chromosome number from their parents and cannot produce viable or fertile offspring when back crossed (Section 2.5). The inception of polyploidy generally does not affect the integrity of an ancestral species' gene pool. So in cases of speciation by polyploidy, ancestral lineages are species on the interbreeding but not species on the phylogenetic approach. Such cases are far from rare given that polyploidy is widespread in plants and often gives rise to new plant species (Mayr 1970, 254–5).

We have yet to see examples in which the ecological and the phylogenetic approaches provide varying classifications. One type of example involves ancestral species. Suppose a new species becomes adapted to an ecological zone different from that of its ancestral species. Furthermore, suppose the ancestral species remains stable in its own niche. In such cases, the ancestral species persists as a single ecological species while on the phylogenetic approach the ancestral species goes extinct (otherwise it would be paraphyletic). Using Figure 4.2, the ecological approach allows that species A persists after the origination of species B, whereas the phylogenetic approach requires that A goes extinct when B becomes a species. Again, how often do such cases occur in nature? Recall that the allopatric model of speciation is thought to have widespread application in nature. Moreover, that model assumes that speciation in part occurs because an incipient species becomes adapted to a new ecological zone. It is that ecological shift, either directly (Paterson) or indirectly (Mayr), that causes a new species to acquire reproductive isolating mechanisms. "[T]he major meaning of reproductive isolation is that it provides protection for a genotype adapted for the utilization of a specific niche" (Mayr 1982, 275).

The above examples amply demonstrate that the interbreeding, ecological, and phylogenetic approaches cross-classify the organic world. At this point, I would like to raise a question. Do the cases cited above provide sufficient evidence for accepting species pluralism? They do not. The history of science is replete with episodes of competing theories in which one theory eventually wins out. The monist could contend that the debate over species approaches is just one of those episodes: current biology contains many competing approaches to species; nevertheless, biologists will eventually determine which one of those concepts (or a future concept) is the right one. Reasons for thinking that pluralism is a more acceptable alternative have not yet been given. The next section provides those reasons in the form of an ontological argument for pluralism. The purpose of this section has been to lay the foundation for that argument by showing that the discrepancy among prominent species approaches is widespread.

4.2 THE ONTOLOGICAL ARGUMENT

What does a monist make of the plurality of approaches to species in the literature? A monist would insist that only one correct approach to species exists and consequently there is only one correct classification of the organic world. Both assumptions will be questioned. As we shall see, current biology provides ample reason for accepting a number of approaches to species. Those approaches provide a plurality of equally legitimate yet varying classifications of the organic world.

It is worth emphasizing that the argument for pluralism offered here is ontological rather than epistemological. Recall the distinction between taxonomic pluralism and metaphysical pluralism in Section 1.4. Taxonomic pluralists maintain that we should allow a plurality of classifications of a single set of entities. But taxonomic pluralists are divided on why we should accept pluralism. Some are motivated by the ontology implied by our best current theories. These taxonomic pluralists are metaphysical pluralists. Other taxonomic pluralists are motivated by what they see as our epistemological limitations. Cartwright (1983) and Rosenberg (1985a, 1994), for example, suggest that because the world is exceedingly complex and we have limited cognitive abilities, we should accept a plurality of simplified yet inaccurate classifications of the world. Though these classifications are inaccurate, they nevertheless answer various scientific needs. Although epistemologically motivated pluralists allow a plurality of classifications, they remain agnostic concerning whether a single correct classification exists. In other words, they are metaphysically agnostic when it comes to pluralism.

The argument in this chapter is that of a metaphysical pluralist. Species pluralism, according to current evolutionary theory, is a real feature of the world and not merely a feature of our lack of information about that world. This is a positive argument and should be distinguished from the view that we are stuck with species pluralism because we do not have sufficient information for choosing among competing species approaches. The positive argument for species pluralism is simply this: according to contemporary biology, each of the three approaches to species highlights a real set of divisions in the organic world. More specifically, evolutionary theory provides the following picture of the organic world. All of the organisms on this planet belong to a single genealogical tree. The forces of evolution segment that tree into a number of different types of lineages, often causing the same organisms to belong to more than one type of lineage. The evolutionary forces at work here include interbreeding, selection, genetic homeostasis, common descent, and developmental canalization (see Templeton 1989 for a discussion of these forces). The resultant lineages include lineages that form interbreeding units, lineages that form ecological units, and lineages that form monophyletic taxa. (Interbreeding units are the result of interbreeding; ecological units are the result of environmental selection; and basal monophyletic taxa owe their existence to common descent.) So the forces of evolution segment the tree of life into a plurality of different classifications: one classification consisting of interbreeding units, another consisting of ecological units, and a third consisting of monophyletic taxa. Of course, this picture of evolution could be wrong. Perhaps some of the above-mentioned forces do not exist, or those forces lack the ability to produce stable taxonomic entities. These are, after all, empirical matters. But given what current evolutionary theory tells us, the forces of evolution segment the tree of life into varying and opposing classifications. Species pluralism is the result of a fecundity of biological forces rather than a paucity of scientific information.

Proponents of monism may allow the existence of different types of base lineages but contend that one type of lineage is more important for understanding the course of evolution. Thus only that type of lineage should be designated by the term "species." Eldredge (1985) and Ghiselin (1987, 1989) provide such an argument in support of the interbreeding approach to species. Eldredge and Ghiselin argue that lineages with sexual organisms are much more important in the course of evolution. Consequently, only interbreeding lineages should be called species. Why do they think that lineages of sexual organisms are more significant? In large part it is because recombination occurs in sexual but not asexual reproduction. In sexual reproduction, genes from both parents are mixed to form the chromosomes of offspring. The resultant chromosomes often contain new combinations of genes that are the source of new phenotypes. In asexual reproduction, there is no mixing of parental genes. Genetic material is passed directly from parent to offspring. Mutation, of course, can alter the genes passed on, but that is true for both asexual and sexual organisms. Recombination causes lineages of sexual organisms to have greater genetic flexibility than lineages of asexual organisms. That flexibility is advantageous in mosaic environments and environments that change over time.

Proponents of the interbreeding approach to species are quick to cite the benefits of recombination in arguing that only lineages of interbreeding organisms should be considered species. For instance, Ghiselin (1989, 74) maintains that asexual organisms do not form species because "clones soon lose out in competition with species. They are very-short-lived, and rarely if ever give rise to significant adaptive radiations." Similarly, Eldredge (1985, 200–1) writes that sexual species "easily outmatch clones and parthenogenetic lineages." He concludes with the following observations:

[S]exually based species simply overwhelm asexual organisms in numbers. As a side effect of sexuality, sexual species are better able to resist extinction than asexual lines. Indeed, strict asexuality (as opposed to some form of alteration of generations) is truly rare and, apparently, always represents a secondary loss of sexuality (ibid.).

Thus, two reasons are given for reserving the term "species" for lineages of sexual organisms. First, groups of sexual organisms have the ability to out-compete groups of asexual organisms. Second, asexuality is rare in nature. Neither reason, I will argue, is sufficient to deny the existence of asexual species.

To begin, why should numbers count? Suppose most of life on this planet reproduces sexually. That in itself does not imply that asexual organisms do not form lineages worthy of being called species. We need a separate argument for that assertion. Putting that aside, it should be noted that the claim that asexuality is rare is a controversial one. Some biologists believe that asexuality is far from rare. Templeton (1989[1992, 164]), for example, writes that asexual organisms constitute "large portions of the organic world." Hull goes further and contends that asexuality rather than sexuality is the prominent mode of reproduction.

When lineages are compared to lineages meiosis is still quite rare, as rare as it should be if it exacts such a high cost. In fact, it turns out to be rare on every measure suggested by evolutionary biologists – number of organisms, biomass, amount of energy transduced, and so on. The only way that it turns out to be common is if sexual and asexual "species" are compared (Hull 1988, 429).

But to do the latter begs the question concerning which lineages form species! According to Templeton and Hull, asexuality is far from rare. Given the disagreement over the occurrence of asexual reproduction, any argument citing the rarity of asexuality to bolster the exclusion of asexual species should be treated with caution.

What of the qualitative claim that recombination allows groups of sexual organisms to out-compete asexual ones? Undoubtedly, recombination is a significant mechanism in sexual organisms. It gives lineages of sexual organisms more genetic flexibility and allows them to adapt more easily to changing environments than asexual ones. Still, asexual reproduction has its advantages, and those advantages have helped maintain the pervasiveness of asexual organisms and their lineages. The process of selfing, for example, gives asexual organisms a leg up in stable environments. Furthermore, selfing is advantageous when conspecific mates are hard to find and when pollinators are scarce (Futuyma 1986, 282). Asexual reproduction is advantageous in some environments, whereas sexual reproduction is better in other environments. Both lineages of asexual organisms and lineages of sexual organisms have done well in the evolution of life on this planet. So the positive reasons that Ghiselin and Eldredge offer for lineages of sexual organisms forming species apply equally to lineages of asexual organisms. Supporters of the interbreeding approach have not provided sufficient grounds for limiting the term "species" to just lineages of sexual organisms.[1]

Still, problems for pluralism lurk elsewhere. Just as some monists object to recognizing asexual lineages as species, others object to recognizing ecological lineages as species. A standard objection is that organisms from different species can and do occupy the same ecological niche (see Wiley 1978[1992], 87; Ghiselin 1987[1992], 376; and Hull 1987, 179). As Hull forcefully poses the problem, "I doubt that a biologist would be willing to put a butterfly and a bird in the same species even if they turned out to be filling the same niche" (Hull 1987, 179). Andersson (1990, 377) considers this sort of objection when applied to rain forest vegetation and responds that a careful look reveals significant ecological differences among species. So perhaps in the end we will find that this objection to the ecological approach is empirically unfounded. Nevertheless, let us allow that on occasion two species do occupy the same niche. Does that spell trouble for the pluralist's use of the ecological approach? No it does not. The above objection is against the ecological approach's universality: it asserts that not all species are differentiated by ecology, so the ecological approach cannot be applied to all species. But the pluralist is not attempting to apply the ecological approach to all species. The pluralist merely holds that the ecological approach is appropriate for some base lineages.

Another objection to the ecological approach points out that the members of a single species can belong to very different niches (see Ghiselin 1987[1992], 376; and Sober 1993, 157). For instance, the dwarf males of some cirripedes live in very different habitats from their conspecific females (Ghiselin, 1987[1992], 376). If the members of the same species live in different habitats, then it seems doubtful that ecological forces are responsible for maintenance of such species. Consequently, the ecological approach does not apply to those species. Ridley (1989, 10) offers a quick response to this objection: "[i]t is no use objecting that different individuals, such as the two sexes, occupy different niches because it would be a trivial statistical exercise, or rotation, to unify the 'different' niches." I am sympathetic to this response. Just as species lineages are complex, multidimensional entities, so are niches. But the pluralist need not be committed to this response. A pluralist can acknowledge that when conspecifics occupy different habitats, forces such as interbreeding or genealogy rather than selection are the dominant stabilizing factor. Again, the pluralist is not committed to a universal application of the ecological approach but merely to the belief that environmental selection is the prominent stabilizing force in some species.

At this juncture, I should again emphasize that the argument for pluralism here is not based on the weakness of each approach to species. I am not asserting that we should adopt various approaches because each is problematic. The argument is quite the opposite: I am highlighting the strengths of the various approaches when applied to different types of lineages. The ecological approach gives us a means for classifying and studying those base lineages in which ecological forces are the primary stabilizing force. The interbreeding approach provides a means for understanding and classifying those lineages maintained by sex. And the phylogenetic approach offers a framework for examining lineages maintained by genealogy. Each species approach highlights an important component of evolution: sex, selection, or genealogy. A biological taxonomy fashioned on only one species approach neglects significant aspects of evolution. In doing so, it provides an impoverished picture of life on this planet.

Thus far I have argued that the tree of life consists of three types of base lineages and that these lineages give rise to distinct classifications of the organic world. One might accept my representation of current evolutionary thinking yet be hopeful that a fourth parameter common to all three types of base lineages will be discovered. Such a parameter would define a fourth type of lineage to which the other types of lineages would be reduced. The result would be a single correct classification of the organic world. The possibility of such an empirical discovery cannot be foreclosed. But a closer look at current biological thinking offers reasons for doubting the existence of such a parameter.

Suppose, for example, one were to suggest that once biologists have performed enough genetic analyses (similar to the current human genome project), they will find that overall genetic similarity is a common parameter of interbreeding, ecological, and monophyletic

143

taxa. This suggestion, however, is implausible. If different species concepts classify the same group of organisms such that one lineage is fully contained in another, then it is impossible that both lineages consist of organisms with the most overall genetic similarity. Recall the example illustrated in Figure 4.1. A is a monophyletic taxon and A + B forms an interbreeding unit. The organisms in A and A + B cannot both have the most overall genetic similarity. For if the organisms in A have the most overall genetic similarity, then the organisms in A + B must have less overall genetic similarity. Given that some base lineages are contained in others, not all base lineages consist of organisms with the most overall genetic similarity.

Other problems stand in the way of reducing interbreeding, phylogenetic, and ecological lineages to lineages with the most overall genetic similarity. Proponents of the interbreeding approach (Mayr 1970, 321–2; Futuyma 1986, 220–2) maintain that the ability of conspecific organisms to reproduce successfully does not turn on those organisms' having the most overall genetic similarity. As Futuyma (1986, 222) and Templeton (1981) note, sometimes reproductive barriers between organisms are caused by genetic differences at just a few loci. In such cases, overall genetic similarity and ability to reproduce successfully may part company.

An analogous problem occurs when attempting to align monophyletic taxa with those having the most overall genetic similarity. Evidence of a monophyletic taxon is the occurrence of derived characters (synapomorphies). Just as reproductive isolating mechanisms may turn on a few loci, so might the synapomorphies that signal a new monophyletic taxon. As a result, two base monophyletic taxa may not be distinguishable on overall genetic similarity but only by a handful of derived characters. Alternatively, consider the case of ancestral species. As we have seen, such species are paraphyletic rather than monophyletic. Nevertheless, the members of some ancestral species have a high degree of genetic similarity (Futuyma 1986, 220). If that is the case, then the properties of having overall genetic similarity and being a monophyletic taxon do not coincide.

Now a reductionist might complain that I have gone about this all wrong. Instead of attempting to reduce the three types of base lineages (interbreeding, ecological, and monophyletic) to lineages with the most overall genetic similarity, I should be looking for some specific genetic factor common to those three types of lineages. As Futuyma (1986, 223) points out, "species owe their existence to specific characters governed by specific genes." But this approach does not provide a common basis for reducing the three types of lineages either. The genes that Futuyma (a proponent of the interbreeding approach) thinks define species are those that affect sexual behavior. Yet in some instances, an alteration in the genetic content of an organism can affect its ecological adaptiveness but not its sexual behavior. For example, being heterozygous rather than homozygous for hemoglobin S in a malarial zone affects one's ecological adaptiveness but not one's choice of mates (Futuyma 1986, 75–6). Furthermore, in instances in which genes controlling morphological distinctiveness and reproductive behavior are separable (Mayr 1970, 322), mutations affecting the former but not the latter can cause the existence of new monophyletic taxa that are not distinct interbreeding units. And in instances in which genetic material governing morphology and ecological behavior are different (Futuyma 1986, 238), mutations affecting the former but not the latter can cause the existence of new monophyletic taxa that are not distinct ecological units. The upshot is that the genetic factors governing the distinctive features of interbreeding, ecological, and monophyletic taxa are separable. Hence, the reduction of these types of lineages to their underlying genetic bases results in three different genetic taxonomies. In other words, the plurality of types of lineages at the macroscopic level is just transferred to a plurality of types of genetic factors at the microscopic level. The attempt to find a common genetic factor that unifies the three types of lineages in question goes against current theory.

The results of this section can be summarized as follows. The forces of evolution produce at least three different types of base lineages (interbreeding, ecological, and phylogenetic) that cross-classify the organic world. Each of these lineages is significant in the evolution of life on this planet. Moreover, there is no fourth parameter to which these types of lineages can be reduced. So according to current biological thinking, the tree of life is segmented into a plurality of equally legitimate classifications. If theory is our best guide to ontology, then we should be metaphysical pluralists concerning species.

4.3 VARIETIES OF PLURALISM

The idea of species pluralism is not new. A number of philosophers and biologists have advocated species pluralism (Ruse 1969, 1987; Dupré 1981, 1993; Mishler and Donoghue 1982; Kitcher 1984a, 1984b, 1987,

The Multiplicity of Nature

1989; Mishler and Brandon 1987; Stanford 1995). This section sketches the problems facing earlier forms of species pluralism. It also indicates how the version of pluralism offered in this chapter avoids those problems.

Ruse (1969, 1987) offers the most conservative form of species pluralism. Ruse acknowledges that different species approaches provide different criteria for sorting organisms. Nevertheless, Ruse makes the following claim: "There are different ways of breaking organisms into groups, and they *coincide*! The genetic species is the morphological species is the reproductively isolated species is the group of common ancestors" (1987, 238). In other words, Ruse believes that a plurality of legitimate species approaches pick out the same set of taxa. He is both a taxonomic pluralist and a metaphysical monist. Ruse's motivation for establishing a coincidence among species approaches is his belief that such a coincidence would indicate the naturalness of species taxa. Following Whewell and Hempel (see Ruse 1987 for references), Ruse takes consilience to be a mark of reality – that is, an indication that a classification is natural rather than artificial. So according to Ruse, if various species approaches citing different biological properties pick out the same taxa, we have good reason to believe that those taxa are real.

The consilience of various species concepts is the ideal that some evolutionists have sought (for example, Mayr 1969, 28). But as illustrated in Sections 4.1. and 4.2, nature has stymied that ideal. Groups of organisms that have the most overall genetic similarity are not necessarily groups of interbreeding organisms (Futuyma 1986, 220; Mayr 1970, 321). Some groups of interbreeding units are not monophyletic taxa (de Queiroz and Donoghue 1988, Rosen 1979). And some groups of organisms that form ecological units are not interbreeding units (Templeton 1989, Grant 1981). This lack of consilience is not limited to a few borderline cases. Consider the case of asexual organisms. As we saw earlier, most organisms in the history of this planet are asexual. As a result, a major discrepancy divides species approaches that recognize both sexual and asexual taxa (the ecological and phylogenetic approaches) and one that recognizes only sexual taxa (the interbreeding approach).[2] Given this lack of consilience, a form of species pluralism that relies on consilience should be rejected.

One other item from Ruse's version of pluralism is worth mentioning. Ruse assumes that the alleged coincidence of species approaches indicates the naturalness of species taxa. However, one message of this

146

chapter is that the naturalness of some entities does not lie at the intersection of various scientific concepts. Realism or "naturalness" does not go hand in hand with unification. We have other reasons to believe in existence of species taxa. For example, Sober (1980[1992], 264) characterizes what is real by what has causal efficacy. Although the interbreeding and phylogenetic approaches do not coincide, it is generally agreed that the forces and lineages they highlight are causally efficacious. So despite our desire for consilience and unification, current biological thinking implies that some corners of the natural world are disunified. (See Dupré 1993 for an elaboration of this theme.)

Mishler and Donoghue (1982) and Mishler and Brandon (1987) offer a more liberal form of species pluralism. Unlike Ruse, they recognize that different species approaches often pick out different types of real species. This can be seen with their phylogenetic species concept. Recall that according to that concept all species must be monophyletic taxa (Section 2.5). Nevertheless, Mishler and Donoghue maintain that "because different factors may be 'most important' in the evolution of different groups, a universal criterion for delimiting fundamental, cohesive evolutionary units does not exist" (1982, 495). Some species taxa owe their existence to reproductive factors, other species taxa are the result of ecological forces, and still others are due to homeostatic mechanisms. Their phylogenetic species concept is monistic in that all species taxa are monophyletic; but it is pluralistic in that different types of processes cause lineages to be species. The result is a single classification of the organic world consisting of different types of basal monophyletic taxa.

There are two problems with Mishler, Donoghue, and Brandon's form of pluralism. First, it requires that all species taxa form monophyletic taxa. Any nonmonophyletic basal taxa that satisfy classic population genetic parameters for specieshood (for example, gene flow and exposure to common selection regimes) must be rejected. But as we saw in Section 4.1, instances abound of paraphyletic base taxa that form stable and significant interbreeding or ecological units. Prime examples of such taxa are ancestral species. Of course biological taxonomists could ignore paraphyletic taxa that form good interbreeding and ecological units, thereby bringing the species category in line with cladism. But doing so results in an overly impoverished classification of the organic world.

The second problem with Mishler, Donoghue, and Brandon's species pluralism is its commitment to a *single* classification of the organic world. As we saw in Section 4.1, different species approaches often cross-classify the same group of organisms. As a result, different species approaches provide varying classifications of the organic world. This variance is not limited to the discrepancy between a classification containing only monophyletic taxa (as in the case of Mishler, Donoghue, and Brandon) and a classification containing both monophyletic and nonmonophyletic taxa (such as one based on the interbreeding approach). Even within a strictly monophyletic classification, there can be monophyletic interbreeding and ecological units that are not coextensive.[3] Because Mishler, Donoghue, and Brandon's pluralism does not allow for the existence of varying but empirically significant classifications, their pluralism does not go far enough.

Kitcher (1984a, 1984b, 1987, 1989) offers a third brand of species pluralism. Kitcher's pluralism differs from that of Mishler, Donoghue, and Brandon in two significant ways. First, Kitcher allows that different legitimate species concepts cross-classify the organic world. Second, Kitcher maintains that some species taxa may not form genealogical lineages. The latter point can be seen with Kitcher's own taxonomy of legitimate species concepts (Figure 4.3). He organizes species concepts into two types: historical and structural. Historical concepts require that species are genealogical entities. The interbreeding, ecological, and phylogenetic approaches are historical concepts: each requires that species taxa form historically (spatiotemporally) continuous entities. The phylogenetic approach corresponds to the first historical concept on Kitcher's diagram, the ecological approach corresponds to the second historical concept, and the interbreeding approach corresponds to the fourth historical concept.[4] Structural species concepts, on the other hand, do not require that species taxa form historically continuous entities. Instead, structural concepts require that the organisms of a species have theoretically important similarities. Kitcher suggests genetic, chromosomal, and developmental similarities.[5]

When looking at Kitcher's taxonomy of legitimate species concepts (Figure 4.3) one cannot help noticing that none of Kitcher's structural concepts are pursued by biological taxonomists whereas many of the historical concepts are. Kitcher's advocacy of structural concepts goes against, or at least goes beyond, current biological practice. Kitcher's point is to stress that biological practice could, and should, allow the legitimacy of historical *and* nonhistorical (structural) species concepts. To make the legitimacy of nonhistorical species concepts intuitive, Kitcher cites a hypothetical case of lizard lineages (1984a, 314–5). The

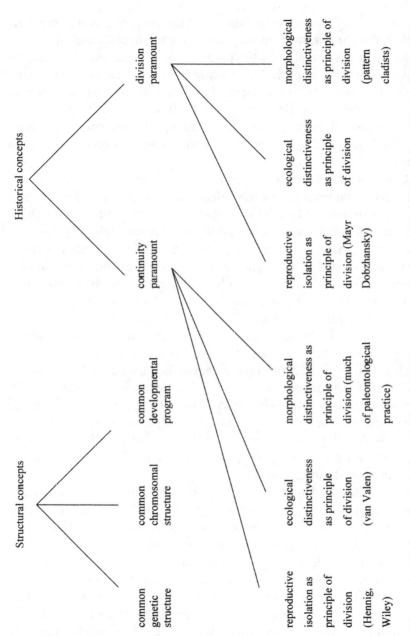

Figure 4.3   Kitcher's organization of different legitimate approaches to species. Adapted from Kitcher 1984a, 325.

lineages are spatiotemporally disconnected from one another; nevertheless, their organisms are very similar along morphological, behavioral, ecological, and genetic parameters. Kitcher writes that "to hypothesize 'sibling species' in this case (and in like cases) seems to me not only to multiply species beyond necessity but also to obfuscate all the biological similarities that matter" (1984a, 315). Kitcher suggests that we allow the existence of spatiotemporally disconnected species taxa and accept the legitimacy of nonhistorical (structural) species concepts.

Unfortunately Kitcher's argument for the legitimacy of nonhistorical species concepts places biological taxonomy outside the range of evolutionary theory. Qualitative approaches to biological classification are out of step with one of the most well-accepted tenets of Darwinism, namely, that a significant way to explain the occurrence of certain traits in a taxon is to highlight that taxon's evolutionary history (Section 3.3). Providing such evolutionary explanations assumes that species and all taxa are genealogical entities. Suppose we offer a selectionist explanation for the prominence of a trait in a taxon. That explanation would cite the origin of the trait and the various selective forces that promote that trait. Selection, or any other biological process, can cause a trait to become prominent among the members of a taxon only if that trait is transmitted by heredity. Hereditary relations, whether they be genetic or otherwise, require that the generations of a taxon be historically connected. The upshot is that if species taxa, or any taxa, are to evolve, they must form historically connected entities. By allowing nonhistorical species concepts, Kitcher's pluralism falls outside the domain of evolutionary biology.

More recently, Kitcher (1989) has offered a further argument for accepting nonhistorical species concepts. As Kitcher points out, asserting that species are historical entities does not sufficiently specify the nature of species, for each organism, population, and all of life on this planet is a historical entity. To understand the nature of species, we need further conditions that distinguish species from other historical entities. Kitcher then shows that the conditions for segmenting the tree of life into species are vague and problematic. Kitcher concludes that we should accept the legitimacy of nonhistorical species concepts (1989, 204). Kitcher is certainly right that the criteria for determining which historical entities are species taxa are controversial, but such controversy in no way nullifies the requirement that species must be historical entities. In more general terms, a condition's insufficiency does not imply that it is not necessary. Species, whether they be basal interbreeding, ecological, or phylogenetic taxa, are historical entities. We are just uncertain about how to draw the boundaries of such taxa. Indeed, such boundaries may be naturally vague (Section 1.3).

Recently, Stanford (1995) has adopted Kitcher's form of pluralism. But while Kitcher is a realist when it comes to species, Stanford believes that species pluralism implies that species taxa do not exist. We have already seen that pluralism does not automatically lead to anti-realism (Sections 1.4 and 4.2). Nevertheless, Stanford thinks that species pluralism does lead to anti-realism. Let's take a quick look at Stanford's argument.

As we have seen, Kitcher advocates accepting a number of species concepts. But Kitcher does not propose that we accept any species concept (see Section 4.4). Kitcher suggests that we accept just those species that stand in "interesting biological relations" (Stanford 1995, 80). Given this suggestion, Stanford provides the following argument for anti-realism.

> [B]y Kitcher's pluralistic criteria, the legitimate interests of biologists *constitute* those divisions recognized as species. Thus, as the course of biological inquiry proceeds, we do not decide that we were previously *mistaken* about which groups of organisms were species; rather, as our explanatory and practical interests change, which divisions of organisms *actually* are species changes as well. As we have noted, which principles of division are biologically interesting can vary without corresponding changes in the world of organisms. Kitcher's species therefore lack that property, supervenience on the state of the mind independent material world, which we demand of real objects (1995, 83).

To illustrate Stanford's argument, consider his example of Cuvier's species concept. According to Stanford, we reject Cuvier's concept not because it provides an inaccurate picture of the organic world, but because it is "wrongheaded" given present-day theory. Biological theory has changed and as a result so have the groups of organisms we recognize as species taxa. The organic world has, more or less, remained constant. Species taxa, therefore, merely reside in our collective theoretical interests.

A significant problem with Stanford's argument is that it is too global. It is not an argument that turns on the adoption of pluralism. Indeed, pluralism is a red herring. Consider an application of Stanford's argument to monism. A monist who adopts realism recognizes that species concepts come and go, but hopes that eventually we will get the correct concept. Stanford's response is that species concepts are only correct relative to a conceptual scheme and not according to the world. So even if one species concept wins out and receives universal acceptance, we should not assume that it highlights mind-independent divisions in nature. Stanford's interpretation of how classifications are chosen is doing all the work here, and that interpretation applies whether one is a monist or a pluralist. Similarly, Stanford's argument applies whether one is talking about species, genes, or electrons. Stanford needs to give more than the standard skeptic's argument against scientific realism. We would like to know why the existence of species is particularly problematic beyond the general claim that the reference of theoretical terms depends on our interests. As it stands, Stanford's argument fails to show why *species pluralism* should cause us to be skeptics. Consequently, Stanford has not provided sufficient grounds for accepting an anti-realist form of species pluralism.[6]

A final form of species pluralism that we will consider is found in Dupré's work (1981, 1993). Dupré (1993, 50–1) argues that "both historical (evolutionary) and structural (or functional) inquiries should be accorded equal weight in biology, and that they may require different classificatory schemes, the latter in some cases demanding a morphological classification." What does Dupré mean by "morphological classification"? In his earlier work (Dupré 1981) he allows that morphological classifications could be based on the phenetic measurement of overall similarity (82–3, 89–90). However, Dupré (1993) has shied away from endorsing nonhistorical classifications using overall similarity. He cites "the difficulties with making sense of absolute similarity" (45). (We will return to this and other problems with the phenetic approach to species in Section 5.5.) Alternatively, Dupré suggests that we accept morphological (i.e., nonhistorical) classifications based on properties that have "suitable evolutionary or other theoretical significance" (ibid.).

Dupré's pluralism is similar to Kitcher's, but Dupré offers his own reason for accepting nonhistorical approaches to species. He acknowledges that evolutionary explanations require that species are historical entities, but he suggests that "species" plays an important role outside of evolutionary theory.

> It is just not the case that the parts of evolutionary theory that deal with events and changes that species undergo is the sole theoretical context

in which species figure. Indeed, there is much more to biology than evolutionary theory. Consider, in particular, ecology. Although ecological theory is presented in terms of fairly abstract categories (prey, predator, parasite, saporphyte, and so on), the application of such a model to a concrete ecological situation will depend on a classification of the organisms involved. Such a classification will draw on general taxonomy so that, for instance, in a particular system lynxes are the predators and hares are the prey. The explananda of such models will be such things as fluctuations of the number in these species. I can see no way of understanding *this* theoretical context other than as requiring the treatment of members of species as members of a kind (1993, 42; also see 20).

The categories of ecological theory that Dupré mentions (prey, predator, etc.) may form theoretically important qualitative kinds, but that fact does not make species taxa nonhistorical entities. In the first place, suppose that hares are the prey of lynxes. Why would that situation render hare a nongenealogical entity? Hares are still members of a particular species taxon because of their genealogical connections. They just happen to be the prey of lynxes as well. Another problem with Dupré's example is the assumption that ecological kinds and species taxa are coextensive. Suppose we want to identify the prey of an insectivorous bird in an ecosystem. The bird may eat insects from hundreds of species and those species may form no higher taxonomic unit. What organisms are a bird's prey depends on what is available and edible, not on a classification according to species and higher taxa. Of course, the prey of a bird may sometimes belong to just one species. But that is an accidental and often temporary feature of the ecosystem being modeled: some edible organisms of different species may wander into the area; the prey belonging to the first species may die because of disease; and so on. The prey of an organism may be a mix of species, and that mix may change over time. The same reasoning applies to the category predator: the predators of an organism may belong to numerous species. Contrary to Dupré's assumption, the kinds of ecology and species taxa need not be coextensive. And even if they were, that coincidence would not show that species taxa are something other than genealogical entities.

Let us summarize the results of this section. Dupré's and Kitcher's forms of pluralism are too liberal because they allow the existence of nonhistorical species taxa. The forms of species pluralism advocated by Mishler, Donoghue, Brandon, and Ruse are not liberal enough – they are committed to the existence of a single correct classification of the world's organisms. The version of species pluralism offered in this chapter charts a middle course between these forms of pluralism. It acknowledges that the forces of evolution create different types of basal taxa. It also recognizes that various types of taxa give rise to classifications that cross-classify the organic world. Still, the version of pluralism advocated here retains the evolutionary insight that species taxa are historical entities. Of the various forms of species pluralism on the market, the one offered in this chapter best captures the virtues of pluralism while avoiding its pitfalls.

4.4   MONIST OBJECTIONS

It should come as no surprise that monists are not very happy with species pluralism. Over the years they have raised a number of objections (Ghiselin 1969, 1987; Sober 1984b; Hull 1987, 1989, 1990; Mayr 1987). One objection was addressed in Section 4.2, namely, that only one type of basal taxon should be accepted as species. But other objections need to be addressed. This section discusses four monist objections. It also raises a methodological challenge to pluralism that the next chapter attempts to meet.

The Communication Objection

Systematists often state that classifications are "information and retrieval systems" (Mayr 1969, 98; Ghiselin 1969, 79; Wiley 1979, 309; Eldredge and Cracraft 1980, 165ff.). For example, Mayr writes that "[a] classification is a communication system, and the best one is that which combines greatest content with greatest ease of information retrieval" (Mayr 1969, 98). With this goal of classification in mind, Ghiselin (1969, 85) provides the following objection to pluralism:

> Whatever standard one does take for ranking taxonomic groups, it should be clear that systematists work at cross purposes when they do not agree on any such criteria. If a common standard were recognized, the system would be more informative by far, and the goal of natural classification would be better served.

Similarly, Hull writes that "[t]erming a hodgepodge of different units 'species' serves no useful purpose as far as I can see. If pluralism entails confusion and ambiguity, I am forced to join with Fodor's . . . Granny in her crusade to stamp out creeping pluralism" (1987, 181). This objection can be codified in the following argument. If we accept species pluralism, then the term "species" will be ambiguous. Such ambiguity will cause confusion in scientific discourse because biologists will talk past one another when they use the term. Such confusion should be avoided. Hence, species pluralism should be avoided. (Additional instances of this argument can be found in Hull 1989, 313; Hull 1990, 84; and Mayr 1987, 165.)

Both Kitcher (1984a, 326–7) and Dupré (1993, 52) offer a quick and appropriate response to this objection: ambiguity need not entail semantic confusion. Oddly enough, that is the case with the current use of "species" in biological taxonomy. As we have seen, biologists define the word differently, and they refer to various types of entities by that term. Biologists are well aware of this ambiguity, and as a result, they often attempt to avoid semantic confusion in their writings by explicitly stating what they mean by "species." In my experience, biologists are sufficiently careful when using the term that the confusion monists fear seldom occurs in the technical literature. The word "species," I would suggest, serves as a counterexample to rather than an instance of the communication objection.

A more radical response to the communication objection suggests entirely eliminating the term "species" from biology. Instead of referring to basal lineages as species, biologists should use terms that more accurately describe various types of basal lineages. Grant (1981, 36), for example, suggests using the terms "biospecies" and "ecospecies" for lineages picked out by the interbreeding and ecological approaches. Add to these the term "phylospecies" for the lineages highlighted by the phylogenetic approach. (Cain 1960, Chapter 7, offers his own set of names for various types of species.) In an earlier writing (Ereshefsky 1992b) I adopted this response to the communication objection. Since then, I have tempered my view. The term "species" is well entrenched in biology and has been used for hundreds of years. School children are taught about species from their earliest encounters with biology, and the word is used in our governments' laws. It is hard to see how "species" could be eliminated from biological and ordinary discourse in the near future. In the meantime, we should exercise care when using the word in technical discussions. To that end, such terms as "ecospecies," "phylospecies," and "biospecies" are useful. They provide a precise way to highlight what is meant by a particular instance of "species." (Further discussion of the fate of the term "species" occurs in Section 8.3.)

The Inconsistency Objection

Some authors reject species pluralism because they believe that it fails to provide a "consistent treatment of the evolutionary process" (Hull 1987, 180; also see Hull 1990, 84). The notion of inconsistency is ambiguous, so we need to consider several ways species pluralism may be inconsistent. Perhaps species pluralism is inconsistent because it allows the existence of different sorts of classifications, and those classifications provide varying representations of the organic world. For instance, when surveying a plot of land we might find that the interbreeding approach identifies three distinct species whereas the phylogenetic approach identifies two species (see examples in Section 4.1). The form of species pluralism presented in this chapter allows such a result. Yet the existence of varying classifications for a particular set of entities does not entail any inconsistency. The form of pluralism advocated here assumes that the tree of life is segmented by different evolutionary forces into several types of basal lineages. Some lineages are maintained primarily by interbreeding, others by ecological forces, and still others by genealogy. The resultant classifications are classifications of *different aspects* of the tree of life that are consistent with one another. Similarly, chemistry provides a consistent treatment of the chemical world even though it sorts entities into different types of classifications – for example, one classification based on molecular structures, another based on phases of matter, a third according to melting temperatures, and so on (see Section 1.4).

Perhaps the charge of inconsistency should be read differently. Species pluralism may lead to inconsistency because it allows that a single organism can belong to two different species. It certainly would be odd to place one organism in two species of the same type – for instance, one organism into two different interbreeding species. The form of species pluralism being presented here is not advocating that option.[7] Something more modest is being proposed. A single organism may belong to multiple species of different *types*. For example, an organism may belong to an interbreeding lineage and a basal monophyletic taxon. There is nothing inconsistent in this view of evolution. It is no more incoherent than my belonging to the class of males and the class of educators, or a soldier being a part of the Canadian military and a part of the United Nations peacekeeping forces.

Yet another way that species pluralism might be inconsistent harks back to the communication objection. Consider a remark by Hennig

(1966, 154): "If systematics is to be a science it must bow to the self-evident requirement that objects to which the same label is given must be comparable in some way" (also see Hull 1990, 84). It may be the case that the species category consists of noncomparable taxa (Ereshefsky 1998). Species consisting of asexual organisms are very different from those consisting of sexual organisms. Different mechanisms cause their unity: at least some sexual species are unified by inter-breeding whereas no asexual species are bound by that process. Furthermore, sexual and asexual species have quite different structures. Sexual species are interactive entities: their populations are causally connected by interbreeding. Asexual species consist of contemporaneous populations that are not causally connected (though they are historically connected to a common ancestor). There are also significant differences between other types of species. Phylogenetic species and paraphyletic interbreeding species have very different structures. The first are monophyletic; the latter are not. The species category seems to be a heterogeneous category of very different types of entities. If that is the case, then Hennig's plea for consistency is violated.

In Part III of this book we will return to the question of whether the species category consists of comparable entities. For now, let us accept, for sake of argument, that the species category contains non-comparable entities. What should we do to avoid this sort of inconsistency? As suggested in the discussion of the communication objection: not much. Dropping the term "species" and replacing it with "phylospecies" and "biospecies" would be an optimal choice in an ideal world. But we do not live in such a world. In the near future we are likely stuck with the term "species" even though it refers to a heterogeneous class of entities. Much more can and should be said about the fate of the species category if indeed it is nothing more than a class of noncomparable entities. We will return to this issue in Chapter 8 when we discuss the future of biological nomenclature.

The "Give It All You Have" Objection

According to Hull (1987, 178), "the only way to find out how adequate a particular conception happens to be is to give it a run for its money. [Retaining] content with a variety of slightly or radically different species might be admirably open-minded and liberal, but it would be destructive of science." Hull, I think, is making two points here. First, pluralists allow a number of accounts of species such that no particular account can be adequately investigated. Second, if scientists fail to analyze and test their concepts, then the empirical success of science will be greatly diminished. Taken together, we have a substantive criticism of pluralism.

This criticism, however, flies in the face of actual taxonomic practice. In the last thirty years, biologists have pursued various approaches to species. Nevertheless, some of those approaches have had a thorough analysis. Consider the extensive work, both conceptual and empirical, done on the interbreeding and phylogenetic approaches. Furthermore, biological taxonomy as a science has not suffered from the investigation of different approaches to species; instead, it has profited by allowing a better understanding of the organic world. An ontological reason for this better understanding may exist. Perhaps the tree of life is divided into different types of basal lineages that play significant roles in the evolution of life on this planet. If that is the case, then various approaches to species have fared well because they capture important facets of the organic world.

One reason that Hull thinks we should pursue only one species approach and "give it a run for its money" is his concern that pluralism will spread our scientific resources too thinly among a number of projects. Hull is certainly on to something here. If a pluralist contends that *every* proposed species approach should get an equal share of funding, then our knowledge of species might grind to a halt: society would need to fund research on all the species concepts discussed in Chapter 2, as well as all species concepts proposed by biologists, philosophers, creationists, and others. This scenario should be feared. But it is not one to which pluralists are committed. As we shall see shortly, and in great detail in Chapter 5, species pluralists can avoid this situation. There is a middle ground between monism and a pluralism that distributes our scientific resources too thinly.

The "Anything Goes" Objection

A number of authors believe that pluralism is an overly liberal approach because it provides no means for discerning legitimate from illegitimate classifications (see Sober 1984b, 334; Hull 1987, 179–80; Hull 1989, 313; Ghiselin 1987, 135–6; and Mayr 1987, 149–50). What, for example, discriminates between classifications based on current evolutionary theory and those based on contemporary creationism or outdated biological theory? The pluralist needs to provide criteria for discerning among proposed species approaches and their associated classifications. Otherwise species pluralism boils down to a position of anything goes.

When looked at more closely, the "anything goes" objection highlights two potential problems with pluralism, one cognitive and the other pragmatic. The first casts species pluralism as an instance of cognitive relativism. Relativists and social constructionists (for example, Barnes and Bloor 1982, and Latour and Woolgar 1986) maintain that no account of nature should be considered true. It does not matter whether those accounts are from contemporary physics or mysticism. The relativist's point is not merely that such accounts may be false, as many previous scientific theories have been. If that were the case, then we could possibly have the right theory in the future. Relativists and social constructionists want to make a much stronger point. We simply lack any rational grounds for asserting that one theory is more likely true than another. Why is that? Ironically, relativists do know something for sure: all accounts of nature stem from our varied social standards and our different personal tastes. Moreover, which theories rise to prominence is a matter of politics and luck. These sources, according to relativists, rule out the possibility of one account of nature having better epistemic credentials than another.

Species pluralists (for example, Kitcher 1987, 190; Dupré 1993, 262–3) clearly do not want to be cast as relativists. Such pluralists believe that we should recognize more than one approach to species; nevertheless, they do not think that we should accept every suggested account of species. The "anything goes" objection comes into play here because it challenges pluralists to do more than just say that a number of species approaches should be accepted. Species pluralists must provide a means for discerning among species concepts; otherwise such pluralism becomes a form of relativism.

The "anything goes" objection cuts yet another way against species pluralism. Here the concern is pragmatic. Society has limited resources for research and education. National granting agencies can fund only a minority of proposed projects. School curricula are constrained by time and money. As a society we must make choices. We would like those choices to be informed, given that today's science education and research have long-term effects on our society. If the pluralist provides no way to discriminate among proposed taxonomic approaches, then the detrimental allocation of resources that Hull voices in the "give it all you have" objection may occur: our science resources will be spread too thinly among a number of taxonomic projects. If that occurs, then no particular approach to classification will be adequately investigated. So for pragmatic reasons, the pluralist needs to provide criteria for discerning which taxonomic approaches should be pursued.

The "anything goes" objection, in both its cognitive and pragmatic form poses an important methodological challenge to pluralists. Species pluralists are well aware of this challenge, and they would like to place limits on pluralism. For example, Kitcher (1987, 190) writes that species pluralism should not condone "*the suggestions of the inexpert, the inane, and the insane*" (his italics). Similarly, Dupré (1993, 51) suggests that pluralism should not "be taken as necessitating complete tolerance of every taxonomic scheme that anyone should happen to think up." Clearly Kitcher and Dupré do not want species pluralism to be a position of anything goes. What, then, do they suggest as a means for discriminating among proposed taxonomic projects?

Kitcher (1984a, 1987) is clear concerning which species concepts he considers legitimate. "Species are sets of organisms related to one another by complicated *biologically interesting relations*" (1984a[1992], 317; italics added). This is a start, but then the question arises, interesting to whom? Kitcher is fairly explicit in answering this question (see 1984a[1992], 323, 328; 1987, 190). "[B]iologically interesting relations" are those highlighted by our major biological theories – evolutionary theory, theories in ecology, microbiology, and elsewhere. Kitcher's way of discerning among taxonomic projects boils down to this: only taxonomic projects proposed by workers using successful theories in well-accepted biological fields should be pursued. Some may be happy with this proposal. Others may not like it but concede that Kitcher has provided the best method possible for judging taxonomic projects. Hull does not fall into either of these groups. He thinks much more should be said about what makes a taxonomic approach a "serious alternative" (1987, 179). Hull writes, "[t]he problem with Kitcher's discussion is that it lacks a sufficiently focused context. The 'current needs of biological research' and the species concept as 'naturalists and theoretical biologists alike use it' are not good enough" (1989, 313).

I am sympathetic to Hull's criticism. I think much more should be said in properly addressing the "anything goes" objection. Even if we restrict the plurality of legitimate species concepts to those that stem from successful biological theories, such a pluralism would still be too liberal. Kitcher's suggestion gives a blank check to any project or concept proposed by a scientist working in a successful theory. Some species concepts might satisfy Kitcher's criterion but should not be pursued because they run counter to the role of species in evolutionary theory. Here I have in mind such concepts as Kitcher's and Dupré's functional and morphological species concepts. Moreover, Kitcher's suggestion for choosing among proposed concepts does not escape the pragmatic side of the "anything goes" objection. We would still have too many projects to fund, and further criteria would be needed to discern among projects that have already passed Kitcher's test. Kitcher's suggestion is a start, but much more needs to done in addressing the "anything goes" objection.[8]

As we saw earlier, Dupré (1993, 51) is another species pluralist who does not advocate "complete tolerance of every taxonomic scheme that anyone should happen to think up." What, then, does Dupré propose as a means for discriminating among proposed taxonomic schemes? In his discussion of species pluralism, he offers a proposal very similar to Kitcher's. Legitimate classifications should be constructed according to properties that have "suitable evolutionary or other theoretical significance" (Dupré 1993, 45). Later he writes that a taxonomic scheme should be accepted if "it serves some significant purpose better than the available alternatives" (1993, 51). Hull's criticism of Kitcher's criteria applies to Dupré's criteria as well. The properties of having "theoretical significance" or serving "some significant purpose" are too broad to determine which species concepts are worth pursuing.

At the very end of his book, Dupré (1993, 242–3, 261ff.) offers a more satisfactory answer to the "anything goes" objection. His concern there is not just species pluralism but more generally how a pluralist should rank and judge proposed research projects. Dupré (1993, 242–3) writes:

> I suggest that we try to replace the kind of epistemology that unites pure descriptivism and scientistic apologetics with something more like a virtue epistemology. There are many possible and actual such virtues: sensitivity to empirical fact, plausible background assumptions, coherence with other things we know, exposure to criticism from the widest variety of sources, and no doubt others. . . . Such an approach would at the very least have the capacity to capture the rich variety of projects of inquiry, without conceding that anything goes.

This is a more robust answer to the "anything goes" objection than merely saying that we should allow only taxonomic projects that serve a "significant purpose" or are grounded in "suitable" theory. The epistemic virtues Dupré suggests offer a means for determining which taxonomic projects and their associated theories are "significant." In addition, they limit which taxonomic projects to pursue when a plethora of projects are "significant" and "suitable." (Recall that even "significant" and "suitable" theories may provide too many taxonomic schemes to pursue.)

Dupré's second answer gets us closer to an adequate response to the "anything goes" objection. However, Dupré does not tell us which of his suggested epistemic virtues should be used in judging proposed projects. Should all of them be applied to every proposed project, or should only a subset be used? Furthermore, should one subset of virtues be used for one discipline, say physics, and another subset be used for a different discipline? Finally, notice that Dupré's list of virtues is an open-ended disjunction: other virtues might be used in judging projects, though Dupré does not explicitly say what they are. In sum, Dupré hints at fruitful ways of answering the "anything goes" objection, but his suggestion is too sketchy. His response is akin to a very incomplete cooking recipe: some ingredients are listed; other necessary ingredients are left unspecified; worse yet, we are not told what to do with the ingredients.

The "anything goes" objection remains a pressing problem for pluralists. Unfortunately, the suggestions of Kitcher and Dupré do not provide sufficient criteria for judging taxonomic projects. The next chapter attempts to provide a fuller answer to the "anything goes" objection. That chapter will take us away from ontological matters to methodological issues in biological taxonomy. This shift to methodology is important. Properly determining whether one should be a pluralist or a monist, or which approaches to classification should be adopted, depends on properly understanding the goals and methods of biological taxonomy. As Kitcher (1984b, 629) notes, biological systematics is a discipline whose "presuppositions need philosophical analysis." Such an analysis is offered in the next chapter. That analysis will take us a long way in answering the "anything goes" objection.

5

How to Be a Discerning Pluralist

The last chapter ended with a challenge to the taxonomic pluralist in the form of the "anything goes" objection. Organisms can be sorted into numerous classifications. We can, for example, sort organisms into species according to breeding behavior, phylogenetic relations, size, color, even which organisms my son likes best. Species pluralists, however, do not propose that we accept every conceivable taxonomic project. Two reasons were offered in the previous chapter. The first reason is a cognitive one. Species pluralists believe that some taxonomic approaches have better epistemic credentials than others and therefore are more worthy of investigation. The business of biological taxonomy, they suggest, is to focus on the most empirically promising approaches. The second reason for restricting pluralism is more pragmatic. Society has only limited resources to dedicate to science and, in particular, to biological taxonomy. National granting agencies can fund only a minority of proposed projects. If pluralism provides no way of discerning among projects, then our scientific resources will be spread too thinly and no project will be investigated properly. The challenge to the pluralist, then, is to allow the investigation of a plurality of classifications while at the same time constraining which classifications should be pursued.

This chapter provides an answer to the "anything goes" objection by providing an account of how one can be a discerning pluralist. Tempered pluralism, the form of pluralism advocated here, provides a substantive middle ground between taxonomic anarchism and taxonomic monism. Tempered pluralism offers an explicit method for evaluating taxonomic approaches. In doing so, it provides a mechanism for limiting which taxonomic approaches should be pursued. Yet unlike monism, tempered pluralism allows that more than one taxonomic approach may be worthy of pursuit.

Before seeing how tempered pluralism works in biological taxonomy, we need to discuss some general methodological issues. A major aim of this chapter is to determine which methodological rules should be used in judging taxonomic approaches. With those rules in hand, we can decide which approaches are worthy of investigation. Examples of candidate methodological rules are "pick only taxonomic approaches that rely on falsifiable principles" and "pick only theory-neutral taxonomic approaches." Before deciding which rules should be used in choosing taxonomic approaches, we need to ask a more general question: How should one select the methodological rules that guide a discipline? Section 5.1 is devoted to answering that question. The approach to methodology adopted in this chapter is Laudan's (1984, 1987, 1990) normative naturalism. Normative naturalism selects methodological rules by first determining the aims of a discipline and then by determining which rules best promote those aims. The remainder of the chapter is an application of normative naturalism to biological taxonomy. Section 5.2 highlights the aim of biological taxonomy, and Section 5.3 offers a list of methodological rules for selecting taxonomic approaches. Finally, Section 5.4 applies the methodological rules derived in Section 5.3 to species concepts. As we shall see, tempered pluralism discriminates among species concepts. Consequently, it provides a concrete answer to the "anything goes" objection.

5.1 NORMATIVE NATURALISM

A standard way to answer the "anything goes" objection is to invoke a universal rule for deciding which theories are scientific and thus worthy of acceptance. Such rules are called "demarcation criteria": they demarcate scientific theories from unscientific ones. We might use a demarcation criterion to discriminate among taxonomic approaches by determining which are scientific. Those deemed scientific should be given funding, and those thought to be unscientific should be abandoned or at least given a low priority. Twentieth-century philosophy of science contains a number of suggested demarcation criteria (see, for example, Popper 1963, Lakatos 1970, Thagard 1980, and Kitcher 1982). A well-known demarcation criterion is Popper's notion of falsifiability. According to Popper, all and only scientific theories are falsifiable. That is, all and only scientific theories can be refuted by empirical evidence. If it is logically impossible to refute the statements of a theory, then that theory is not scientific. Some cladists cite Popper's criterion of falsifiability in arguing for their taxonomic approach (see Eldredge and Cracraft 1980, 67; Wiley 1981, 111; and Nelson and Platnick 1981, 47–8, 339ff. for a detailed discussion of the use of falsifiability in biological taxonomy see Sober 1988, Section 4.2).

A growing view in the philosophy of science is that a universal demarcation for science does not exist (Laudan 1983, Rosenberg 1985b, Dupré 1993). The problem, they suggest, is not our lack of insight into the nature of science. The problem is the multifaceted nature of science. Scientific disciplines are so varied that no single criterion can adequately characterize all of science. To take just one example, Popper's notion of falsifiability omits large chunks of science. Consider such probabilistic statements as "Radon atoms have a probability of .5 of decaying during any period of 3.82 days." Science is rife with probabilistic statements, yet such statements are unfalsifiable. Probabilistic statements are unfalsifiable because it is impossible to prove conclusively, using empirical evidence, that they are false. No finite number of radon atoms failing to decay within 3.82 days shows that the above probabilistic statement is false. The above statement asserts that radon atoms have a certain probability of decaying during that period; it does not assert that they *will* decay. Even if no radon atoms decay, the probabilistic statement can be true. (Of course, the statement would then be unlikely, given the evidence. But still, that evidence does not prove that the statement is false, as falsifiability requires.) Much of science consists of probabilistic statements, so much of science is deemed unscientific by Popper's criterion. Popper's notion of falsifiability, therefore, fails as a universal demarcation criterion for science.

Popper's criterion is just one of many suggested demarcation criteria. This is not the place to introduce various criteria for science and rehearse the arguments against their universality (Laudan 1983, Rosenberg 1985b, and Dupré 1993 are good sources for that project). Instead, I merely want to raise the point that if we are going to look for methodological rules for discriminating among taxonomic approaches, we would be wise not to look for a universal criterion for science. The search for *the* demarcation criterion for all of science is fraught with difficulties. As many writers have observed, the aims of different scientific disciplines vary, and as a result, so do their methodological rules (see, for example, Laudan 1984, Gould 1986, and Hacking 1996). A more fruitful way to answer the "anything goes" objection is to determine which methodological rules pick out theories that best satisfy the aims of a specific discipline. In particular, we should answer the "anything goes" objection to species pluralism by determining the aims of biological taxonomy and then deciding which methodological rules select species concepts that satisfy those aims. Laudan's (1984, 1987, 1990) normative naturalism offers such an approach to methodology.

Before turning to normative naturalism, I should point out that even if we had an adequate demarcation criterion we would not have a proper answer to the "anything goes" objection. The history of science is full of old taxonomic approaches that are rejected by contemporary scientists. Such approaches may pass a demarcation criterion – they may indeed be scientific – but that does not imply that they should be investigated. A graphic example is a classification based on Aristotelian physics. Such a classification is part of our scientific heritage; however, our current empirical knowledge provides strong reasons for rejecting it. In this case, and others like it, we use criteria beyond demarcation criteria for determining whether a project should be pursued. Similar remarks apply to debates in biology concerning currently proposed taxonomic approaches. Scientists argue over which taxonomic approaches should be pursued while at the same time allowing that competing approaches may be equally scientific. Again, even if we had an adequate demarcation criterion, we would be left with the question of which scientific taxonomic approaches should be pursued.

As suggested above, a more fruitful way of answering the "anything goes" objection can be found in Laudan's (1984, 1987, 1990) normative naturalism.[1] Normative naturalism selects methodological criteria for evaluating taxonomic approaches without relying on a universal criterion for science. According to Laudan, disciplines (such as biological taxonomy or particle physics) contain three major components: general aims, methodological rules, and empirical theories. The general aims of a discipline are its overriding cognitive goals. Examples of general aims include the desire for infallible knowledge, the desire for realistic theories, and the desire for theories that allow us to predict certain phenomena. Methodological rules help us pick which theories are worthy of further investigation. Methodological rules are sometimes called "methodological criteria" or "epistemic virtues." We have already seen many examples of such rules. They include "prefer theories that make surprising predictions," "prefer simpler theories," "reject inconsistent theories," and "adopt only falsifiable theories."

Our main interest in normative naturalism is how it selects methodological rules. According to normative naturalism, we should prefer those methodological rules that best gauge a theory's ability to achieve the aims of a discipline. For example, suppose a general aim of a discipline is to expand our empirical knowledge of certain phenomena. Suppose, also, that theories not vulnerable to empirical evidence more often than not provide little in the way of new empirical knowledge. In that situation, normative naturalism counsels the scientists in that discipline to adopt the rule "reject theories that are not vulnerable to empirical evidence."

Methodological rules, according to normative naturalism, have no intrinsic worth by themselves but serve as means for achieving a discipline's general aims. Laudan (1987) likens methodological rules to hypothetical imperatives of the form "if your aim is e, you ought to do m." For example, the methodological rule "reject theories that are not vulnerable to empirical evidence" is merely an elliptical version of the hypothetical imperative "if your aim is to provide theories that expand empirical knowledge, then you ought to reject theories that are not vulnerable to empirical evidence." Methodological rules, thus, are tools that guide us in picking theories that promote specific general aims. Notice that normative naturalism is a local approach to methodology rather than a global approach. It selects methodological rules for a particular discipline according to the aims of *that* discipline. As a result, normative naturalism may advocate different methodological rules for different disciplines. The set of methodological rules highlighted in Section 5.3, for instance, is for biological taxonomy; that set may or may not apply to other disciplines.

How, then, do we know which methodological rules best gauge a theory's ability to achieve a discipline's aims? According to normative naturalism, there are several ways. One is to look at the history of science and see which methods have best promoted a specific aim in the past. For example, Laudan (1984, 127ff.) conducts a survey to determine whether the aim of acquiring empirically successful theories is best promoted by the rule "prefer theories that entail earlier theories." He argues that such a rule is not warranted: Copernican astronomy did not retain the key mechanisms of Ptolemaic astronomy; and relativistic physics did not retain the mechanisms of aether theory. So, according to Laudan, physicists should not adopt the methodological rule "accept only theories that entail earlier theories" if their aim is to achieve empirically successful theories.

We can now see why normative naturalism is naturalistic. Often philosophical arguments concerning methodological rules involve just conceptual analysis or a priori reasoning. According to normative naturalism, empirical considerations should also affect our choice of methodological rules. One way to judge methodological rules is to conduct an empirical investigation using episodes in the history of science as data. Laudan's historical survey of the rule "accept only theories that entail earlier theories" is an instance of this approach. Another way to evaluate methodological rules is to use episodes from current science. Those rules that promote our desired aims in contemporary science should be adopted; those rules that do not promote such aims should be rejected. Finally, empirical information from science itself (rather than the history or sociology of science) affects our choice of methodological rules. For example, the preference for theories tested by randomly collected data (versus nonrandomly collected data) stems from empirical theories (Laudan 1984, 38). We have learned that the entities we encounter often are not representative of all the entities in the world. So one methodological rule we adopt is "prefer theories tested by randomly collected data (when possible) over theories tested by nonrandomly collected data."

This discussion shows the various ways that information about the empirical world can affect our choice of methodological rules. Normative naturalism, however, does not limit debate over such rules to just empirical evidence. Conceptual analysis also applies in these debates. Think of an analogous situation in science. Often debates concerning competing theories cannot be settled by empirical data, so conceptual considerations come into play. The same is true in choosing methodological rules. If empirical matters do not indicate which rules best promote a discipline's aims, then conceptual analysis must be employed. As Laudan (1990, 50–1) writes:

> I am not claiming that the theory of methodology is a wholly empirical activity, any more than I would claim that theoretical physics was a wholly empirical activity. Both make extensive use of conceptual analysis as well as empirical results.

Normative naturalism is naturalistic because its methodology is no different from that of the sciences. Moreover, normative naturalism's choices are affected by the results of science. On this approach to methodology, the divide between philosophy of science and theoretical science is far from sharp.

Thus far we have seen how normative naturalism selects method-ological rules. What remains to be seen is how it determines a discipline's aims. In the simplest form, it is just a matter of reading such aims from the discipline itself. Take a survey of the leading texts and articles of a discipline and see what their authors advocate as the aims of that discipline. If there is agreement on those aims, plug them into the apparatus of normative naturalism and determine the appropriate methodological rules. However, a worry may arise: Do the workers in any discipline agree on their discipline's aims? Rosenberg (1990) and Leplin (1990) believe that such agreement does exist. They suggest that the aim of *all* scientific disciplines is the acquisition of empirical knowl-edge. I am not ready to assume that all disciplines have the same general aim, let alone assume that such agreement exists within a par-ticular discipline. First we need to survey scientists and see what they take to be the aims of their respective disciplines. A survey of the leading texts of a discipline may reveal that there is a lack of consen-sus concerning that discipline's aims. Using the apparatus of normative naturalism that diversity of aims might lead to different sets of method-ological rules. In the next section I will suggest that this problem does not arise for biological taxonomy. Prominent taxonomists tend to agree on the general aim of their discipline. As a result, biological taxonomy is an ideal field for the application of normative naturalism. Neverthe-less, if we were dealing with a discipline with diverse and contradictory general aims, normative naturalism does provide mechanisms for evaluating such aims.

One way to discriminate among proposed aims, according to Laudan (1984, 50–3), is to reject those aims that are unrealizable. Laudan cites the quest for infallible empirical knowledge as an unrealizable aim. His reasoning is none other than an application of Hume's problem of induction: the universal generalizations found in many empirical theo-ries apply to more instances than can ever be checked; consequently, it is impossible to know with certainty that such generalizations are true. Another way to evaluate general aims is to see if they cohere with a discipline's acknowledged "canonical achievements" (Laudan 1984, 53ff.; 1990). Suppose a group of scientists disagree over the general aims of their discipline. Nevertheless, they agree upon the canonical achievements of that discipline. If one set of aims is consistent with those achievements while another set is not, then we should prefer the former set of aims. For example, physicists in the 1700s and early 1800s disagreed on a general aim of their field (Laudan 1984, 55–9). Some desired theories that posit only observable entities; others preferred theories that explain a broad range of phenomena, regardless of whether those theories posit unobservable entities. In the later part of the eighteenth century, both parties agreed that the best theories at the time posited unobservable entities. As a result, physicists in the beginning of the nineteenth century tended to adopt the aim that allowed theories to posit unobservable entities. These scientists, according to Laudan, evaluated their aims by how well those aims agreed with the canonical achievements of their discipline.

It is time to step back from the details of normative naturalism and apply it to the "anything goes" objection. Such an application comes in two stages. First, we determine the aims of the discipline in question – here biological taxonomy; then we select methodological rules that best promote those aims. The next two sections are devoted to those tasks. With the appropriate methodological rules in hand, a pluralist can choose among taxonomic projects and be a discerning pluralist.

5.2   THE AIM OF BIOLOGICAL TAXONOMY

To gain a better understanding of the general aims and methodological rules of biological taxonomy, I turn to a plausible set of authorities on the matter – leading proponents of the four major schools of biological taxonomy. What follows is a survey of seminal texts and articles in that field. Prominent workers from each of the four major schools (evolutionary taxonomy, phenetics, process cladism, and pattern cladism) are canvassed. As we shall see, explicit statements of the general aim of biological taxonomy are not hard to find. They appear in influential papers and in the beginning chapters of seminal texts.

A good place to start is with the school of evolutionary taxonomy and one of its major proponents, Ernst Mayr. In his text on the history of biology, Mayr suggests that biological taxonomy has two primary goals. The "practical" aim of a biological taxonomy is to provide classifications that tell us which taxa belong to more inclusive taxa. A classification, according to Mayr, "should serve as an index to an information storage and retrieval system" (1982, 148). The general aim of biological taxonomy is different:

> Since the observations in all comparative branches of biology were organized with the help of "the natural system" (now evolutionarily defined),

it became *the primary function* of classification to delimit taxa and construct a hierarchy of higher taxa which permitted the greatest number of valid generalizations (1982, 149, emphasis added).

Clearly Mayr believes that one takes precedence, namely, that classifications should allow the production of "valid generalizations." Further evidence that Mayr takes this to be the overriding aim of biological taxonomy is found in his article, "Biological Classification: Toward a Synthesis of Opposing Methodologies" (1981). There he writes, "[a] far more important function of a classification, even though largely compatible with the informational one, is that it establishes groupings about which generalizations can be made" (1981[1994], 279).

What sorts of generalizations does Mayr have in mind, and why does he think they are so important? The generalizations he alludes to tell us which taxa belong to more inclusive taxa and which characteristics may be found among the members of a taxon. Such generalizations are important because they serve as a basis for making various types of inferences about the organic world. Examples of such inferences include inferring the evolutionary history of a taxon, inferring the relatives of a taxon, and inferring what characters may be found in related taxa (Mayr 1969, 8–9). With such inferences in hand, biologists can explain and predict various biological phenomena. Here is an illustration from Mayr's text on biological systematics (*The Principles of Systematic Zoology*):

> One of the greatest assets of a sound classification is its predictive value. It permits extrapolation from known to previously unstudied characters. An analysis of a few species strategically scattered through the natural system may provide us with much needed information on the distribution of a new enzyme, hormone or metabolic pathway. Many animals cannot be kept in the laboratory, and others will not reproduce in captivity. Again, a sound system will permit all sorts of inferences from the genetically well-known types (1969, 7).

Similar statements by evolutionary taxonomists are found in Simpson (1961, 25) and Crowson (1970, 12–14). Leading evolutionary taxonomists agree that the overall aim of their discipline is to construct classifications that serve as a basis for drawing inferences about the organic world.

Let us now turn to phenetics and its major proponents, Sokal and Sneath. In his paper, "The Continuing Search for Order" (1985), Sokal suggests three primary purposes for constructing biological classifications. The first is to provide a general reference system that "assigns an organism to a taxon, gives it a name, associates it with other related taxa . . . , and enables statements and inferences to be made about the characteristics of a given group" (1985, 737–8). "A second purpose is to generate hypotheses about evolutionary relationships" (1985, 738). Sokal's examples of such hypotheses concern the number of species found in a genus or the density with which species fill niche spaces. "A third purpose of a classification is to serve as a model or benchmark against which to test hypotheses about evolutionary phenomena, such as phyletic evolution, gradualism versus punctuated-equilibrium evolution" (1985, 738). In each of these "primary purposes" of biological taxonomy, classifications are seen as hypotheses (or generalizations) from which biologists can make various inferences. Some inferences are about the characters found within a taxon; others concern the tempo and mode of evolution.

At the level of general aims, Mayr and Sokal are not that far apart. They both desire classifications that provide hypotheses that can be used for making inferences. Such inferences, they agree, can be about the sorts of characters one finds within a taxon, or more general ones such as the pattern of speciation. Of course, important differences separate pheneticists and evolutionary taxonomists (Chapter 2). But those differences do not occur at the level of general aims. Pheneticists and evolutionary taxonomists, I would suggest, agree on the general aim of biological taxonomy yet disagree on which methodological rules best achieve that aim. Consider an important difference between pheneticists and evolutionary taxonomists. Pheneticists attempt to produce theory-neutral or operational classifications, while evolutionary taxonomists rely on evolutionary theory in producing classifications. Despite this difference, authors on both sides are clear that they prefer theory-neutral or theoretically based taxonomic approaches *because* such approaches serve as means for producing more reliable classifications. In other words, they disagree on the means for achieving the ultimate aim of biologically taxonomy, but not on that aim. Consider Sokal's (1985, 737–8) argument for adopting a theory-neutral approach to classification. He first lists "the purposes of classification," namely, to provide various sorts of hypotheses concerning the organic world. He *then* argues that a theory-neutral approach is the best way to achieve those purposes (1985, 738ff.). Similar lines of reasoning are found in Sokal and Sneath (1963, 55) and Sokal and Crovello (1970[1992], 51). Here we have a clear distinction between general aims and the methodological rules that are used to achieve those aims.

I have suggested that proponents of evolutionary taxonomy and phenetics agree on the ultimate aim of biological taxonomy (though they disagree widely on how to achieve that aim). It is time to turn to process cladism and the founder of cladism, Willi Hennig. In perhaps his most general statement of the aims of biological taxonomy, Hennig (1966, 7) writes, "We have characterized the tasks of systematics as the investigation and presentation of all relations that exist among living natural objects, and systematics as all scientific activity that aims at ordering and rationalizing the world of phenomena." For Hennig, classifications are "reference systems" and the task of systematics is to determine various ways of constructing reference systems. The major focus of Hennig's remarks in the introductory chapter of his *Phylogenetic Systematics* (1966) is to determine which reference system (morphological, phylogenetic, ecological, and so on) should serve as *the* general reference system for all biological disciplines (1966, 7, 9). As we saw in Chapter 2, Hennig argues that the phylogenetic system is the preferred system.

Eldredge and Cracraft (1980) and Wiley (1981) have written prominent texts on process cladism and articulate Hennig's goals for biological systematics in the familiar terminology of "hypotheses" and "inferences." For instance, Eldredge and Cracraft (1980, 20; also see 5) write:

> In this book, we are concerned primarily with erecting hypotheses about the pattern of life's history, including the supposed sequence of phylogenetic branching events and the distribution of similarities and differences among the organisms being studied. . . . These hypotheses, then, should attempt to explain, as simply as possible, all the relevant empirical data bearing on the questions of what has been the history of life.

In other words, biological taxonomy should provide us with hypotheses concerning genealogical relations and the distributions of certain characters. These hypotheses can then be used to explain (or at least infer) the history of life.

Very similar remarks on the aim of biological taxonomy are found in Wiley's text (1981). In the introductory chapter, Wiley mentions the various interests of comparative biologists – for example, the study of predator-prey interactions, and the study of genotypic change during speciation. He points out that "[w]hatever the interest, the comparative biologist can work most efficiently if he or she has a reference system to consult which will help to pick critical comparisons" (1981, 15). For Wiley, reference systems (i.e., classifications) provide hypotheses concerning phylogenetic relations, and those hypotheses are needed to study such phenomena as genotypic change during speciation because they tell us which genotypes are ancestral and which have undergone speciation. Again, the aim of taxonomy is to provide hypotheses from which biologists can draw inferences.

Our survey is not yet complete. The other branch of cladism, pattern cladism, needs to be considered. Recall that pattern cladists adopt Hennig's practice of constructing classifications using only synapomorphies. Yet they reject Hennig's evolutionary justification concerning which characters are synapomorphies and why those sorts of characters should be used. Pattern cladists preserve the pattern portion of cladism – classifications reflect nested groups of organisms as evidenced by synapomorphies. But they ignore the processes that give rise to those patterns. Process cladists, including Hennig, attend to both cladistic patterns and their underlying processes.

Despite these important differences, pattern cladists espouse a similar ultimate aim of taxonomy. Consider the following quotation from the primary text on pattern cladism, Nelson and Platnick's *Systematics and Biogeography* (1981):

> [C]lassifications obviously perform an essential function in information storage and retrieval. They allow us to deal with tremendous amounts of data by subsuming a great deal of information into single words. . . . But is this task of data organization *the* essential function of classification? If it were, would we not, as scientists, leave the task of constructing classifications, of organizing our huge pile of data, to technicians, just as we leave the task of organizing our huge pile of publications to librarians? Why, then, do scientists concern themselves with constructing classifications? Perhaps because classifications serve not only to summarize the information we already have, but also to predict information that we do not yet have. For example, if there is a set of organisms, all of which share properties (A, B, C, D, E) that no other organisms have, and we find another organism about which we know only that it has properties A and B, can we not predict that it will have properties C, D, and E as well? . . .
>
> Classifications, then, are useful not merely as data summaries but also as hypotheses about order in nature. These hypotheses, once tested and corroborated, can be used in studies of aspects of nature other than the attributes of organisms. It is as hypotheses that classifications are ultimately useful (9).

Again, the aim of biological taxonomy is to produce hypotheses that allow biologists to infer characters within a taxon and to infer more general features of the organic world (that is, "aspects of nature other than the attributes of organisms").

Although pattern and process cladists agree on the overall aim of biological systematics, they part company on how to achieve that aim. It is their disagreement over methodological rules that divides pattern and process cladists. Nelson and Platnick (1981) believe that synapomorphy should be "viewed in purely empirical terms" (165), thereby avoiding "futile theorizing" (151). Hennig, however, takes the opposite view: "nothing at all is achieved by a completely nontheoretical ordering of organisms. All results of biological investigation . . . have meaning only if a realm of applicability can be seen in them that extends beyond the individuals (single organisms) from which they were derived" (1981, 8). For Hennig, classifications have meaning and applicability only if they are couched within theories about process. For pattern cladists, theory dependence should be avoided at all costs. Both are talking about what sorts of information should be included in classifications so a classification can serve as a meaningful basis for making inferences. Again, they are not disagreeing over the aim of biological taxonomy, but over how classifications should be constructed to achieve that aim.

Let us take stock of what has been accomplished so far. Leading proponents of the four major schools of biological taxonomy have been surveyed. Surprisingly, the taxonomists surveyed agree on the overall aim of their discipline, though they disagree on how best to achieve that aim. That aim, as we have seen, is to provide empirically accurate classifications that allow biologists to make inferences. The information provided in such classifications is necessary for making various inferences about the organic world. Empirically accurate classifications help biologists infer the evolutionary history of a taxon, the close relatives of a taxon, what characters are typically found among the members of a taxon, even the mode and tempo of evolution. Empirically accurate classifications provide the backbone of biological theorizing. It is no wonder that taxonomists place the attainment of such classifications as their discipline's highest goal.

One might object to the above conclusion along the following lines. Sure, biological taxonomists of all stripes want to produce classifications that allow biologists to draw various sorts of inferences. Nevertheless, those taxonomists disagree on the types of inferences that classifications should help provide. This division shows that the various schools of classification have different overall aims.[2] For my part, I do not think that the desire for different types of inferences indicates that workers in the four schools disagree on the aim of their discipline. Just as process cladists and evolutionary taxonomists want to construct hypotheses about evolutionary history, pattern cladists and pheneticists allow that their hypotheses may be seen as evolutionary. For example, Nelson and Platnick (1981, 199) suggest that "cladograms, and branching diagrams generally, may be viewed from an evolutionary perspective, and viewed as such they may be termed phyletic trees." And Sokal (1985, 746) writes, "for those whose major purpose is to estimate phylogenies, phenograms would in many cases give estimates of the true cladogeny." On the other hand, both evolutionary taxonomists and process cladists observe that one type of hypothesis derivable from classifications concerns the characters found among the organisms of a taxon (see Mayr 1969, 7, and Eldredge and Cracraft 1980, 7; both of these passages are quoted earlier in this section).

A further possible concern with this survey is how the division between aims and methodological rules is drawn. What evidence is there for claiming that the desire for various kinds of inferences is the aim of taxonomists while the remaining criteria are methodological rules for promoting that aim? More generally, what reasons can one produce for showing that some criteria are a discipline's aims while other criteria are its methodological rules? Laudan (1984, 1987, 1990) does not address this issue directly. I suspect that he takes the distinction between the aims and methodological rules to be intuitively obvious. That may be. Nevertheless, a close look at the writings of the biologists cited in this section offers some concrete reasons for drawing the aim/methodological rule distinction as I have.

Recall that according to normative naturalism, methodological rules are considered derivative of a discipline's aims – they are rules that serve as means for selecting projects that best satisfy those aims. When one turns to the methodological writings of leading taxonomists, one finds very much the same structure: a clear distinction is made concerning which criteria are the general aims of biological taxonomy and which criteria serve as means (methodological rules) for achieving those aims. For example, Sokal (1985) offers a description of the "purposes of taxonomies" immediately followed by "optimality criteria" for judging which general taxonomic schools can best accomplish the purposes of biological taxonomy (737–8). Similarly, Nelson and Platnick's

(1981) introductory chapter contains a section titled "Methods in Comparative Biology" immediately followed by a section called the "Purposes of Classification." Additional examples of this structure of argumentation can be found in the introductory chapters of Hennig (1966), Eldredge and Cracraft (1980), and Wiley (1981).

Furthermore, biological taxonomists often explicitly state what they take to be the overriding aim of their discipline. According to Mayr, *"the primary function of classification* is to delimit taxa and construct a hierarchy of higher taxa which permit[s] the greatest number of valid generalizations" (1982, 149, emphasis added). Another example is from Eldredge and Cracraft (1980): "[i]n this book, *we are concerned primarily* with erecting hypotheses about the pattern of life's history, including the supposed sequence of phylogenetic branching events and the distribution of similarities and differences among the organisms being studied" (20, emphasis added). Additional examples of biological taxonomists explicitly stating the aim of their discipline can be found in the quotations offered earlier in this section.

Yet another possible concern with the survey is the vagueness of the suggested aim of biological taxonomy. One might complain that consensus among taxonomists concerning the aim of biological taxonomy is achieved only by providing an overly vague formulation of that aim. The intuition here is that I have cheated in deriving a single of goal of biological taxonomy by vaguely articulating that goal. I agree that the stated goal of biological taxonomy is at a high level of generalization. However, that the suggested aim of biological taxonomy is quite general and not as precise as one might want does not show that it is not the aim of that discipline. We have seen numerous quotations from leading biological taxonomists stating that this is the aim of their discipline. We can criticize the aim they offer, but we would be wrong in asserting that it is not the aim they espouse.[3]

Before leaving this section, I would like to emphasize that though leading biological taxonomists agree on the aim of their discipline, they widely disagree on which methodological rules best promote that aim. We have already seen that pheneticists and pattern cladists believe that classifications should be theory neutral with respect to evolutionary theory, whereas evolutionary taxonomists and process cladists believe that classifications should be theory dependent. Further methodological disagreements are easily found in biological taxonomy. Mayr (1981, 290), for example, argues that classifications should be general and encompass various biological parameters. Thus he promotes classifications that capture information concerning common ancestry and evolutionary divergence. Hennig, on the other hand, prefers classifications based on a single uniform parameter: "If systematics is to be a science, it must bow to the self-evident requirement that objects to which the same label is given must be comparable in some way" (1966, 154; also see 22). Consequently, he promotes a taxonomic approach based on the sole parameter of common ancestry.

Another example comes from the species debate. As we saw in Section 2.5, the evolutionary species concept highlights groups of organisms that have their "own evolutionary tendencies and historical fate." Unique tendencies and fates are caused by several processes, including interbreeding, selection, and genetic homeostasis. Proponents of the interbreeding approach to species argue that their approach is preferable because it provides a uniform account of species: the interbreeding approach posits one process to explain the unity of species while the evolutionary species concept posits several (see Mayr 1987, 165, and Ghiselin 1987, 138). Proponents of the evolutionary species concept respond that their approach is preferable because it is more general: the evolutionary species concept applies to both sexual and asexual organisms, whereas the interbreeding approach applies only to sexual ones (see Simpson 1961, 153–4; and Wiley 1978[1992], 82).

We have spent a considerable amount of time surveying the methodological writings of biological taxonomists. The results of that survey point to a single overall aim of biological taxonomy. That aim can now be plugged into the apparatus of normative naturalism to answer the "anything goes" objection. Recall that according to normative naturalism, theories (or in our case, taxonomic approaches) are chosen by a discipline's methodological rules. Those rules are selected by their ability to pick theories that promote the aims of a discipline. We have the aim of biological taxonomy in front of us. So we can now turn to an analysis of which rules select taxonomic approaches that promote that aim. With those rules in hand, a discerning pluralist can say which species concepts should be accepted.

5.3 PRIMARY AND SECONDARY RULES

As we saw in Section 5.1, normative naturalism offers several ways to select methodological rules. One is to consult the history of science to discover which methodological rules have promoted an aim in the past.

Another is to use conceptual analysis to arrive at which methodological rules would promote that aim. A third approach lets empirical information from science affect the selection of methodological rules. In this section, the latter two approaches are employed. That is, conceptual analysis and empirical information from science are used to derive methodological rules for selecting taxonomic approaches in biological taxonomy.[4]

Before evaluating which rules should be used in judging taxonomic approaches, some terminology should be clarified. "Taxonomic approach" is a phrase used throughout this chapter. It refers to specific approaches to constructing classifications. The four general schools of biological taxonomy are examples of taxonomic approaches; so are the various species concepts. A taxonomic approach produces a classification by using two types of principles – "sorting principles" and "motivating principles." Sorting principles sort entities into taxonomic units. Motivating principles justify the use of sorting principles. As an example, consider the biological species concept. Its sorting principles assert: sort organisms that can interbreed and produce fertile offspring into the same species, sort organisms that reproduce sexually but cannot successfully interbreed into different species, and sort organisms that reproduce asexually into no species. The motivating principle of that approach assumes that groups defined by interbreeding relations form stable evolutionary units.

Let us now turn to the examination of methodological rules. Biologists and philosophers suggest numerous rules for judging taxonomic approaches. I divide those rules into two classes: primary rules and secondary rules. Primary rules best gauge whether a taxonomic approach can satisfy the aim of biological taxonomy. They serve as minimal though fallible standards for judging a taxonomic approach. Secondary rules, on the other hand, affect a taxonomic approach's value but are less important in judging whether that approach will satisfy a discipline's aim. In what follows, I suggest that four rules best gauge a taxonomic approach's ability to satisfy the aim of biological taxonomy. As a result, they should be considered primary rules. Those rules are empirical sensitivity, internal consistency, intertheoretic coherence, and intratheoretic coherence. Why these four rules and not others? I consider each rule in turn.

First, every biological taxonomist surveyed writes that the primary aim of the discipline is to provide *empirically accurate* classifications from which biologists can make various types of inferences. A minimal way of judging whether a taxonomic approach can provide empirically accurate classifications is to see if its sorting and motivating principles are sensitive to empirical evidence. For example, if the sorting principles of a taxonomic approach sort organisms according to their niches, then we should be able measure niches. If niches cannot be measured, then there is no way to judge the empirical accuracy of that approach's classifications. The same consideration applies to motivating principles. A number of species concepts are motivated by the assumption that interbreeding, and only interbreeding, causes clusters of organisms to be stable evolutionary units. If we cannot test that assumption, then we cannot judge whether classifications should or should not contain species of asexual organisms.

The criterion of empirical sensitivity should not be confused with verifiability or falsifiability. Verifiability assumes that empirical evidence can demonstrate the truth of a hypothesis. But as Hume's problem of induction shows, many empirical hypotheses cannot be proven with certainty. The only way to prove the truth of the hypothesis "All H_2O boils at 100 degrees Celsius" is to heat *all* instances of H_2O to 100 degrees Celsius and see whether they boil. Needless to say, such an experiment cannot be performed. For this reason, as well as others, Popper suggests his criterion of falsifiability: all scientific hypotheses must be capable of being proven false by empirical evidence. As we saw in Section 5.1, however, many scientific hypotheses (for instance, probabilistic ones) cannot be proven false. In contrast to verifiability and falsifiability, the criterion of empirical sensitivity merely requires that empirical data can affect the probability we assign to a hypothesis. The philosophical literature abounds with probabilistic accounts of the relation between evidence and hypothesis.[5] Although the notion of empirical sensitivity needs refinement, it stands as an important desideratum for taxonomic approaches.

Second, if classifications are to fulfill the aim of providing a basis for inference, they need to be relatively unambiguous. Suppose we want to infer the phylogenetic history of a taxon so that we can understand the physiology of its organisms. A taxonomic approach infected with ambiguity may place that taxon in different higher taxa. Those higher taxa may have distinct evolutionary histories. Consequently, we would not be able to explain the current physiology of the organisms in question because we would have multiple and inconsistent explanations. For example, Gould (1980) explains why today's pandas have a false thumb fashioned from the wrist's radial sesamoid bone by citing what happened to their ancestors. If a taxonomic approach places pandas in multiple higher taxa (of the same rank), and those taxa have been exposed to different evolutionary forces, then we may produce several inconsistent explanations of the panda's thumb.

One might wonder if the species pluralism advocated in Chapter 4 conflicts with the rule of internal consistency. Perhaps some forms of species pluralism lead to conflicting classifications that then result in conflicting explanations. However, the form of species pluralism advocated in Chapter 4 does not send us down that road. Different species approaches classify organisms according to different empirical parameters. If we want to know the genotypic differences among species of interbreeding organisms, we consult classifications based on interbreeding. Alternatively, if we want to know the genotypic differences among species maintained by ecological forces, we consult the ecological approach to species. There is no conflict here. What the criterion of internal consistency prevents is the production of multiple classifications according to a single parameter, not the production of multiple classifications according to different parameters.

Third, motivating and sorting principles, like all scientific hypotheses, are accepted given certain background assumptions. Sorting principles that rely on recombinant DNA techniques for constructing phylogenetic classifications depend on tenets in physics, chemistry, and biochemistry (Panchen 1992, 207). The plausibility of the biological species concept's motivating assumption that interbreeding causes evolutionary unity depends on tenets in genetics (Templeton 1989, 166–8). We would like the background assumptions that bolster sorting and motivating principles to cohere with what is known elsewhere. When a taxonomic approach conflicts with well-established theories elsewhere we should be wary. Of course, this is not a foolproof method. A taxonomic approach may make an assumption that conflicts with a well-accepted tenet elsewhere, yet it may prove more empirically fruitful in the end. Nevertheless, intertheoretic consistency provides a way to identify suggested taxonomic approaches that have a low probability of empirical success given everything else known at the time. One cannot ask more of a methodological standard than that.

Fourth, the motivating principles of a taxonomic approach should cohere with the current theory governing the entities being classified. The intention of this standard is to prevent inconsistency between a taxonomic approach and a well-established theory in the same field. The history of biology is strewn with taxonomic approaches containing motivating principles that conflict with current evolution-ary theory. As a result, we do not fund or teach such approaches other than as history of science. Just as we are skeptical of outdated taxo-nomic approaches, we should be skeptical of currently proposed bio-logical classifications that conflict with contemporary evolutionary theory.[6]

Having suggested the above primary rules for judging taxonomic approaches, a question arises: Must a taxonomic approach satisfy each of these criteria to be considered worthy of pursuit? A taxonomic approach that satisfies all four rules is worthy of pursuit, and an approach that satisfies none should be investigated only as a last measure. The more of these rules an approach satisfies, the more reasons we have to pursue it. I am hesitant to draw a more precise boundary. The above standards are offered as plausible guides for choosing taxonomic approaches, not as necessary and infallible guides. Some may be unhappy with the vagueness of this proposal, but the exis-tence of vagueness around a putative distinction does not show that the distinction is not viable. Consider analogous cases. Bald and not bald, rich and not rich, even nitrogen and oxygen are vague distinctions (see Sober 1980, 253 on the last distinction). Yet because there are clear cases on either side of these distinctions, we accept them as workable. Similarly, the distinction between taxonomic approaches that are worth pursuing and those that are not is workable if there are clear cases on either side. As we shall see in Section 5.4, the primary rules clearly dis-tinguish taxonomic approaches in biology.[7]

The rules discussed do not exhaust the methodological rules offered by biologists and philosophers. We have not, for example, considered such rules as generality, simplicity, and stability. Such rules, I will suggest, are secondary in judging whether a taxonomic approach can satisfy the aim of biological taxonomy. Pragmatic reasons exist for pre-ferring general, simple, and stable classifications. But these qualities are less vital than the primary rules in judging whether a taxonomic approach can provide empirically accurate classifications. The attain-ment of the qualities highlighted by secondary rules depends on coop-eration from the empirical world. Yet the biological world is often uncooperative.

Consider the virtue of generality. Generality is useful for organizing information about the world. The more general a law, the more we can explain with it; the more general our laws, the fewer laws we need to know. The same is true of biological taxonomy. The more general a taxonomic approach, the fewer taxonomic approaches biologists need to know. Whether biological taxonomy ultimately will contain a single general approach depends on cooperation from the organic world. As the arguments of Chapter 4 indicate, the organic world may be stymieing that ambition. According to current evolutionary biology, the organic world consists of three types of basal lineages: interbreeding, ecological, and monophyletic. No single approach to species captures all three of these types of units. Interbreeding concepts leave out stable and distinctive lineages of asexual organisms. Ecological concepts leave out interbreeding lineages whose members occupy different niches. And phylogenetic concepts neglect interbreeding and ecological lineages that are paraphyletic. So when it comes to species, the desire for classifications that allow us to make theoretically important inferences conflicts with the desire for a single general classification. Given that conflict, the virtue of generality should not be taken as a primary rule for selecting taxonomic approaches.

A similar point applies to the virtue of simplicity. According to Wiley (1981, 20), "the principle of simplicity (parsimony) is used to pick the hypothesis that explains the data in the most economical manner." The interbreeding approach to species is simpler than the phylogenetic approach. The interbreeding approach explains the existence of species taxa by positing one process (interbreeding), while the phylogenetic approach cites several (interbreeding, ecological forces, and genetic homeostasis). The greater complexity of the phylogenetic approach, however, does not make it any less desirable than the interbreeding approach. Numerous groups of organisms that biologists would like to draw inferences about fail to be taxa on the interbreeding approach (for example, stable groups of asexual organisms and stable groups of geographically isolated sexual organisms). The phylogenetic and not the interbreeding approach offers the resources for drawing inferences concerning such groups. As a result, we should ignore the greater complexity of the phylogenetic approach and accept it on the grounds that it better accounts for the phenomena at hand.

The above example illustrates why simplicity should not be taken as a primary rule for judging taxonomic approaches. Nonetheless, the above example does not show that simplicity should never be used as a criterion for judging taxonomic approaches. Consider the case in which two approaches are equally supported by empirical evidence and both approaches allow us to draw the same types of inferences. In addition, suppose one approach is simpler and its greater simplicity renders that approach an easier one to use. In that case we have pragmatic reasons for preferring one approach over the other. But notice that the criterion of simplicity comes into play only after both approaches score equally high on empirical accuracy and intertheoretic coherence. Simplicity, therefore, is a secondary rather than a primary rule for judging taxonomic approaches.

With generality and simplicity, taxonomists would like classifications to be relatively stable (Michener 1964, de Queiroz and Gauthier 1992, Brummitt 1997b). Frequent taxonomic revision should be avoided not only because such revision is itself inconvenient but also because it requires biologists to relearn classifications (Wiley 1979, 325–6). Again, the possibility of stable classifications turns on cooperation from the empirical world. Unfortunately, our limited access to the organic world frequently causes biologists to revise classifications. New data can cause a taxonomist to divide a previously recognized taxon, lump together what were thought to be different taxa, or shift a taxon from one higher taxon to another. For example, discoveries concerning the reproductive behavior of two groups of organisms may indicate that what was once thought to be two species is in fact one interbreeding species. Alternatively, a new DNA analysis or new fossil discovery may reveal that a genus placed in one tribe should be placed in another. Unlike chemistry, wholesale taxonomic revision is the norm rather than the exception in biological classification. (Specific examples of taxonomic revision are given in Chapter 6.) Of course, stability is a desirable feature of classification. It is a feature that Wiley's annotated Linnaean system and various non-Linnaean systems of classification strive to attain (see Chapters 7 and 8). But stability should not be taken as a primary rule for judging taxonomic approaches, given that our access to the organic world often renders biological classifications unstable. Simply put, stability places too stringent a test on taxonomic approaches.

A further problem with using stability as a primary rule is that the mere stability of a classification does not indicate whether that classification provides a basis for making theoretically useful inferences. Consider a taxonomic approach that constructs classifications according to when biologists discover taxa. Such classifications would be quite stable. No new evidence would cause their revision other than the knowledge that our clocks were wrong. Despite the stability offered by such a taxonomic approach, its adoption would not promote the general aim of biological taxonomy. The parameter it uses for constructing classifications has nothing to do with the organic world beyond us. So very few, if any, theoretically interesting inferences about the organic world can be drawn from such classifications.

We now have two sets of methodological rules in front of us. Empirical sensitivity, internal consistency, intertheoretic coherence, and intratheoretic coherence serve as primary rules for judging taxonomic approaches. Such rules best gauge whether a taxonomic approach can satisfy the aim of biological taxonomy. They serve as minimal though not foolproof standards for judging taxonomic approaches. Generality, simplicity, and stability, on the other hand, are less effective in determining whether a taxonomic approach will be able to satisfy the aim of biological classification. As a result, they are secondary guides for deciding whether a taxonomic approach should be pursued. Secondary rules are nevertheless important. They highlight pragmatic, albeit secondary, reasons for preferring certain taxonomic approaches. Furthermore, the differential preference biologists place on secondary rules motivates the production of new and fruitful taxonomic approaches. (We will consider this latter point in Section 5.5.)

I have tried to motivate the division between primary and secondary rules by discussing the relationships between various methodological rules and the general aim of biological taxonomy. My tactic has been to discern which rules would best judge a taxonomic approach's ability to satisfy the overall goal of biological taxonomy. In large part, my arguments have been in the abstract: I have argued which rules *would* best pick taxonomic approaches. It would be nice to see if the primary rules do in fact distinguish taxonomic approaches. That is, it would be desirable to move from a theoretical argument concerning the applicability of the primary rules to an actual application of those rules. The next section does just that.

5.4   A CASE STUDY: SPECIES CONCEPTS

In this section, the primary rules are applied to four contemporary species concepts. If the primary rules clearly rank those concepts, then we have a case study attesting to the feasibility of employing those rules. Before embarking on this study, several caveats should be mentioned. First, the survey of species concepts offered here is not meant to exhaust the current species concepts in the literature. There are many more species concepts in the biological literature than the four discussed here (see Section 2.5). For the purposes of this survey, we need only a handful of concepts to show that the primary rules can discriminate among taxonomic approaches. Second, the following case study does not attempt to provide a comprehensive analysis of the species concepts discussed. Again, the point of this section is merely to illustrate that the primary rules serve as an effective means for choosing taxonomic approaches. Third, I have chosen species concepts as the focus of the case study rather than the four schools of biological taxonomy. The focus on species concepts reflects the main motivation for this chapter, namely, to provide an answer to the "anything goes" objection to species pluralism. If the primary rules allow us to discriminate among species concepts, then the "anything goes" objection to species pluralism will be answered. The primary rules could be applied to the four general schools. Indeed, the four schools are, in an indirect way, under scrutiny here. Each of the four species concepts reviewed represents one of the four major schools, and the motivating and sorting rules of those schools are often incorporated in their respective species concepts. Nevertheless, the focus here will be species concepts, not the general schools.

Let us start with the phenetic species concept.[8] A common charge is that the phenetic species concept, and phenetic taxonomy in general, fails to provide a single classification for a given set of organisms (Hull 1970, Ridley 1986). In other words, the phenetic species concept is thought to violate the rule of internal consistency. Here are two reasons. First, each organism has an exceedingly large number of characters – too many characters to record for phenetic classification. A pheneticist must choose one set of characters and see which organisms have the most overall similarity according to those characters. This choice of characters is important. As pheneticists themselves state, different sets of characters may give rise to different groups of organisms having the most overall similarity. That, in turn, causes the production of multiple phenetic classifications for the same collection of organisms. Unfortunately, the phenetic species concept cannot provide a way to select which set of characters to use in constructing a classification. This is not a technical or epistemic problem that might be overcome in the future. In principle, the phenetic species concept cannot provide a way to discern among multiple phenetic classifications of a group of organisms. Recall that one of pheneticism's guiding methodological rules is to consider all characters on a par in order to avoid a theoretical bias toward one set of characters. So for the pheneticist, any set of characters must be considered as good as another in constructing a classification (assuming that those sets have the same number of characters). The result is the production of multiple classifications with no basis for preferring one classification over another.

The phenetic species concept gives rise to taxonomic ambiguity at a different level. Suppose we have a set of characters for a group of organisms. A pheneticist plots this information on a graph. Each dimension represents the occurrence of a character, and organisms are plotted according to whether a character is present (for an example, see figure 2.7 and accompanying discussion). A pheneticist then measures the "distance" among organisms on the graph. That distance is a measure of the overall similarity among organisms. Ambiguity arises because there are different ways of calculating overall similarity (Ridley, 1986, 36ff.). For instance, one type of distance measures overall similarity as a function of nearest neighbors on a graph. Another measures overall similarity as a function of the nearest furthest neighbors. As pheneticists themselves show, different distance measures result in varying classifications of a single set of organisms.

The phenetic species concept rates poorly on the criterion of internal consistency. How does it fare on the criteria of empirical sensitivity, intratheoretic coherence, and intertheoretic coherence? I know of no conflict between the phenetic species concept and intertheoretic consistency. Concerning empirical sensitivity, Sokal (1985) compares phenetics and cladism using a hypothetical data base with a known phylogeny. He concludes that under certain circumstances (when the data contain few characters or many taxonomic units), phenetics provides a better estimate of true phylogeny than cladistics. So Sokal believes that phenetics is empirically sensitive. A number of critics of phenetics seem to agree when they argue that phenetics should be rejected because overall similarity and empirical estimates of phylogeny often part company (for example, see Ridley 1986, Chapter 2; and de Queiroz 1988).

That leaves the criterion of intratheoretic coherence. Pheneticists maintain that one reason for adopting a theory-neutral approach using overall similarity is that such an approach best captures the results of evolution (see Sokal and Crovello (1970[1992], 50–1). But some classifications based on strictly phenetic principles conflict with well-accepted assumptions in evolutionary biology. As we saw in Section 3.2, a purely phenetic approach to classification places the sexes of a single species into two different species. Doing so conflicts with the convention that sexual organisms belong to species containing members of both sexes. A purely phenetic approach even sorts the different temporal stages of a single organism into two different species. This outcome goes against the generally accepted assumption that an organism belongs to a species for its entire life. The phenetic species concept, therefore, fails the criterion of intratheoretic coherence. Putting these results together, we see that the phenetic species concept passes the criteria of empirical sensitivity and intertheoretic consistency, but fails to satisfy the criteria of internal consistency and intratheoretic coherence.

The second species concept I would like to consider is from pattern cladism. Pattern cladists, like pheneticists, attempt to provide theory neutral approaches to classification. We just saw that the pheneticist's commitment to theory neutrality causes pheneticism to violate the criterion of internal consistency. The pattern cladist's commitment to theory neutrality gives rise to the same problem (Ridley 1986, 89ff; Sober 1993, 182). Pattern cladists, like all cladists, use synapomorphies as evidence for constructing classifications. But in order to use such evidence a cladist must first distinguish synapomorphies from those characters that fail to serve as cladistic evidence. The process cladist distinguishes synapomorphies from other sorts of characters by making assumptions about the process of evolution: characters change during evolution and synapomorphies are derived character states that have changed from earlier, ancestral states (Ridley, 1986, 54). The pattern cladist cannot make use of such assumptions because to do so would require making theoretical assumptions about evolution. Pattern cladists are in the same boat as pheneticists: any character is as good as any other in constructing classifications. Pattern cladism, thus, fails to provide a means for choosing opposing classifications and is open to the charge of taxonomic ambiguity.

Problems for a pattern cladist approach to species loom elsewhere. Despite pattern cladists' explicit commitment to theory neutrality, their species concept conflicts with contemporary evolutionary theory (Beatty 1982). Consider what Davis and Nixon say about their pattern cladist definition of species, which they call "the phylogenetic species concept (PSC)."[9]

In its requirement that a phylogenetic species be "diagnosable by a unique combination of character states," the PSC stipulates that every comparable individual of a phylogenetic species carry the entire complement of characters of that species. In contrast, if an attribute is fixed in one population but is a trait in another, or if two populations share a trait and differ only in its frequency of occurrence, or if an attribute is unique to one population but is not fixed, this attribute cannot distinguish phylogenetic species (1992, 427–8).

This certainly sounds like a return to essentialism. (Similar quotes are found in Nelson and Platnick 1981, 324, 328, 304.) Yet as we saw in Section 3.1, essentialism is at odds with evolutionary theory. So the pattern cladist approach to species fails the rule of intratheoretic consistency.

How does the pattern cladist species concept fare on the other two primary rules for evaluating taxonomic approaches? I am not aware of any conflict between pattern cladism and well-accepted tenets outside of evolutionary biology. That leaves empirical sensitivity. Nelson and Platnick write that one reason for accepting pattern cladism is that it relies on empirical fact: "The concept 'patterns within patterns' seems . . . an empirical generalization. . . . The concept rests on the same empirical basis as all other taxonomic systems (the observed similarities and differences of organisms)" (1981, 141–2). Whether nested patterns of synapomorphies can simply be observed without theoretical assumptions is questionable. Nevertheless, Nelson and Platnick seem to allow that we can test for nested patterns of synapomorphies. The pattern cladist species concept, therefore, passes the criterion of empirical sensitivity. It also satisfies the criterion of intertheoretic coherence. However, it fails on the virtues of internal consistency and intratheoretic coherence.

The next concept in our survey is Mayr's biological species concept. That concept is a prime example of an evolutionary taxonomist's species concept. It also serves as a representative for other interbreeding-based concepts (for example, concepts offered by Ghiselin 1974 and Patterson 1985 [1992]). Let's start with the criterion of empirical sensitivity. Both supporters and detractors of the biological species concept believe that the motivating principle of that concept – that gene flow is an important causal factor in maintaining stable groups of organisms – is testable. Mayr, for instance, claims that he has empirical evidence supporting his concept (1970, Chapters 5 and 6; 1988, 318–20). Detractors of the biological species concept argue that they have falsifying evidence. They argue that empirical data show that interbreeding is neither necessary nor sufficient for the unity of many species (Ehrlich and Raven 1969, Donoghue 1985, Templeton 1989). The biological species concept passes the criterion of empirical sensitivity with flying colors.

What about the criteria of intratheoretic and intertheoretic coherence? I know of no reason to think that the biological species concept is inconsistent with well-accepted theories outside of evolutionary biology. When we turn to intratheoretic coherence, things become more controversial. As Hull (1987, 179) observes, a strong assumption among taxonomists is that all organisms belong to species. The biological species concept may not satisfy that assumption, or even the weaker assumption that most organisms belong to species. As we saw earlier, asexual organisms do not form species on the biological species concept; yet asexuality may be the predominant form of reproduction on this planet (see Sections 3.4 and 4.2). So the biological species concept's exclusion of asexual species renders it inconsistent with the assumption that all or most organisms belong to species. Mayr (1987, 165–6) himself recognizes this problem and is troubled by it.[10] The problem may extend beyond asexual organisms – many sexual organisms may fail to belong to species according to the biological species concept. As we saw in Section 3.4, some biologists think that many species of sexual organisms consist of populations that exchange little or no genetic material through gene flow (Ehrlich and Raven 1969, Levin and Kerster 1974, Templeton 1989). Others suggest that even when gene flow is present, it is not responsible for the cohesiveness of many sexual species (Endler 1973; Grant 1980, 167). The claim that interbreeding fails to be a strong stabilizing force in many sexual species is controversial. Still, the case of asexual organisms indicates that the biological species concept may conflict with the well-accepted assumption that all or most organisms belong to species.[11]

The remaining primary rule is that of internal consistency. Some (for example, Kitcher 1989 and David Magnus [personal communication]) suggest that the biological species concept gives rise to ambiguous classifications because the distinction between gene flow within a species and hybridization across species is vague. Undoubtedly this distinction is vague. An instance of introgression may signal the fusion of two species, or it may merely be an instance of hybridization. It all depends on whether such interbreeding eventually leads to the integration of two gene pools. Such borderline cases, however, do not give rise to the sort of taxonomic ambiguity one finds in the phenetic species concept. Given the tenets of the phenetic species concept, every case is one in which we can have multiple classifications: we just need to use a new set of similarities to get another classification (see the discussion of the phenetic species concept earlier). Cases in which the biological species concept cannot determine whether we have one or two species certainly occur, but the existence of such cases does not tell against the biological species concept: no species concept can give a determinate classification in all cases. This outcome should be expected, given that speciation, on any respectable account of species, is a process that often has borderline cases. The rule of internal consistency is not intended to guard against the existence of such borderline cases. Indeed it should not. The rule of internal consistency is intended to guard against taxonomic approaches, like the phenetic species concept, that can never provide an unambiguous classification of a set of organisms. In sum, the biological species concept satisfies the criteria of internal consistency, empirical sensitivity, and intertheoretic coherence. But it is problematic when it comes to the rule of intratheoretic coherence.

The final concept I would like to consider is Mishler and Brandon's (1987) phylogenetic species concept. Their concept is a process cladist's approach to species. An important motivating assumption of Mishler and Brandon's concept is that other processes besides interbreeding can cause groups of organisms to form stable evolutionary units. That assumption is vulnerable to empirical evidence: one can test whether asexual organisms or isolated groups of sexual organisms form cohesive entities as the result of processes other than interbreeding. Some of the debate over the phylogenetic approach to species centers on this empirical question (see, for example, Ghiselin 1989). So the phylogenetic species concept passes the criterion of empirical sensitivity. Turning to the criterion of intertheoretic coherence, I know of no conflict between the phylogenetic species concept and well-established tenets in other empirical disciplines. What about the criterion of internal consistency? Undoubtedly there are borderline cases for which the phylogenetic species concept is unable to determine a specific classification for a group of organisms (Mishler and Brandon 1987 even discuss such cases). But as we saw in the discussion of the biological species concept, such borderline cases do not violate the rule of internal consistency.

The remaining rule is that of intratheoretic consistency. Here some controversy exists. The phylogenetic species concept requires that all species are monophyletic (Mishler and Brandon 1987, 46). As we saw

191

in Section 4.1, this rules out the existence of any species whose members give rise to a new species. An example will help to illustrate this point. Suppose a species has a peripheral population that gives rise to a new species. The parental species will not contain all and only the descendants of a common ancestor: some of its descendants will be members of the offspring species. In such cases, the parental species fails to be monophyletic and thus fails to be a species on the phylogenetic species concept. As Hull (1988, 139–40) notes, Hennig had pragmatic reasons for not allowing the existence of ancestral (parental) species. But Hull adds that "[s]ome species *have* to be ancestral" (ibid.). The phylogenetic species concept seems to conflict with the standard evolutionary assumption that ancestral species exist, regardless of whether cladists have good pragmatic reasons for denying their existence (Sober 1988, 21; Ereshefsky 1989).[12] So the phylogenetic species concept of Mishler and Brandon is problematic with respect to the rule of intratheoretic coherence. Nevertheless, it passes the rules of internal consistency, empirical sensitivity, and intertheoretic coherence.

This completes our survey of four contemporary species concepts. As mentioned at the start, this survey is not offered as a complete analysis of the concepts discussed but merely as a demonstration of how one might employ the four primary rules for evaluating taxonomic approaches. The results of the survey are summarized in Table 5.1.

As mentioned in Note 7, the satisfaction of a rule comes in degrees. The biological species concept and the phylogenetic species concept conflict with some assumptions in evolutionary biology, but not as grossly as the other two concepts surveyed. Accordingly, I have labeled the biological and phylogenetic species concepts as "problematic" with respect to intratheoretic consistency, whereas the other two concepts are given a designation of "fail."

The results of this survey suggest a clear division among species concepts. The phenetic and pattern cladist concepts each fail two rules, and the biological and phylogenetic concepts marginally fail one rule each. That division provides fairly straightforward guidance concerning which concepts should be preferred. If we want to pursue those concepts that have the best chance of achieving the aim of biological taxonomy (given all that we now know), we should prefer the biological and phylogenetic concepts. The phenetic and pattern cladistic concepts fall in the second tier. Notice that the primary rules do not discern between the biological and phylogenetic concepts: both are ranked

Table 5.1. *Results of survey of four contemporary species concepts*

Species Concept	Empirical Sensitivity	Internal Consistency	Intratheoretic Coherence	Intertheoretic Coherence
Phenetic	Pass	Fail	Fail	Pass
Pattern Cladist	Pass	Fail	Fail	Pass
Biological	Pass	Pass	Problematic	Pass
Phylogenetic	Pass	Pass	Problematic	Pass

equally. Consequently, we have good reason to pursue both simultaneously. The primary rules counsel taxonomic pluralism when it comes to species, though a pluralism of a very discerning sort.

5.5 METHODOLOGICAL UNITY AND DISUNITY

We have discussed a wide range of issues in this chapter. Let us summarize what has been accomplished. The chapter began with the "anything goes" objection to species pluralism. In an effort to answer that objection, I showed how one can be a pluralist while at the same time limiting which taxonomic approaches should be considered legitimate. There are various ways to do this; I have used normative naturalism. Normative naturalism recommends first determining the aim of a discipline, then deriving which methodological rules promote that aim. Such a procedure was applied to biological taxonomy: leading workers in biological taxonomy were surveyed concerning the aim of their discipline, then rules for promoting that aim were derived. Some rules, which I have dubbed "primary rules," provide a way to filter proposed taxonomic projects. Such rules allow the pluralist to be a discerning pluralist, and consequently, allow the pluralist to answer the "anything goes" objection. The last section is an example of tempered pluralism in action: the primary rules derived in this chapter do discriminate among species concepts.

Primary rules perform a second function. They underwrite the plausibility of pluralism. If multiple projects score equally high on the primary rules, then all of those projects stand an equal chance of satisfying the aim of biological taxonomy. In such a situation we ought to adopt taxonomic pluralism, for in such a situation there is no way to discern among taxonomic approaches. Notice that the question of whether taxonomic pluralism translates into metaphysical pluralism is a separate issue. Perhaps the world is carved in multiple ways, each corresponding to a particular taxonomic approach (that was the thesis of Chapter 4). On the other hand, perhaps in time only one taxonomic approach will be shown empirically accurate. Nothing I have said in this chapter forecloses the possibility of metaphysical monism. Either way, given a situation in which several approaches score equally high, we should adopt a pluralist stance and pursue more than one approach.

Let us turn to the secondary rules. Secondary rules serve a different function from primary ones. As we saw in Section 5.2, biologists prefer varying secondary rules. Some prefer uniformity over generality, others prefer generality over uniformity, still others prefer stability over generality. This differential commitment to secondary rules motivates new and important avenues of research in biological taxonomy. That research, in turn, causes biologists to promote different taxonomic approaches. In other words, the differential commitment biologists have to secondary rules is in part responsible for taxonomic pluralism. Consider an example. The desire for general taxonomic approaches has motivated some biologists to offer new species concepts. Simpson (1961, 153–4), Wiley (1981, 26, 35), and Templeton (1989[1992], 164, 170) argue that a major defect of the interbreeding approach to species is its limited range of application. They would prefer a concept that applies to both sexual and asexual organisms. Accordingly, they offer species concepts that allow that species are maintained by a number of evolutionary forces (interbreeding, selection, and genetic homeostasis). Proponents of the interbreeding approach, on the other hand, resist such alternative species concepts in large part because such alternatives fail to provide a single, uniform account of species. In other words, proponents of the interbreeding approach prefer uniformity over generality. Ghiselin (1987, 372–4), for instance, argues that species concepts that recognize asexual species provide an inconsistent view of evolution: they conflate different types of entities (sexual organisms and asexual organisms) as well as different types of processes (interbreeding, selection, and genetic homeostasis) under one concept. Ghiselin prefers a more uniform treatment of the organic world, particularly the one offered by the interbreeding approach. A number of other writers cite the virtue of uniformity in their support of the interbreeding approach. See Dobzhansky (1951, 274–5), Mayr (1987, 165) and Hull (1987, 168–83).

From the above example, we can see that the differential commitment biologists have to various secondary rules motivates different avenues of research. Further examples of this sort are easy to find in biological taxonomy. As we saw in Sections 2.3 and 5.3, one of Hennig's (1966) motivations for developing cladism was his desire for a single, uniform approach to biological classification. Hence, Hennig suggests classifying according to the single parameter of common ancestry. Mayr (1981) disagrees. He defends evolutionary taxonomy by pointing out that it has the virtue of being more general – it captures both common ancestry and evolutionary divergence. In the chapters that follow, we will see other instances of secondary rules motivating different taxonomic approaches. In the debate over biological nomenclature, proponents of non-Linnaean systems (for example, Griffiths 1976) are motivated by the desire for a uniform system of nomenclature. Defenders of the Linnaean system (for example, Wiley 1979), on the other hand, believe that biologists should continue using the Linnaean system in order to promote stability. (The details of this debate are presented in Part III.) In general, secondary rules are valuable in motivating new and fruitful avenues of research.

Putting the primary and secondary rules together, we can see how the interplay between methodology and taxonomy gives rise to pluralism in biological classification. Secondary rules motivate a plurality of classifications. Primary rules filter which classifications should be pursued while allowing the existence of a plurality of legitimate classifications. I do not claim to have the definitive argument for which methodological rules should serve as primary ones for judging taxonomic projects. Perhaps further work will reveal that the division between primary and secondary rules should be drawn differently. Nevertheless, this chapter amply shows that we have the methodological resources for answering the "anything goes" objection to species pluralism.

III

Hierarchies and Nomenclature

6

The Evolution of the Linnaean Hierarchy

Throughout this book we have looked at issues framed in terms of the Linnaean hierarchy. Does membership in a *species* depend on the qualitative properties of its organisms or the causal relations among them? What distinguishes species taxa from such higher taxa as *genera, families*, and *orders*? Should we adopt monism or pluralism when it comes to the *species* category? Thus far we have taken the Linnaean system for granted. But should we?

Before addressing that question, it should be noted that the Linnaean system of classification is much more than a hierarchy of taxonomic ranks. Linnaeus also provided rules for sorting organisms into taxa, as well as rules for naming taxa. These features of the Linnaean system were premised on Linnaeus's biological theory, in particular, his assumptions of creationism and essentialism. Needless to say, these assumptions have gone by the wayside in biology and have been replaced by evolutionary theory. Still, the vast majority of biologists use the Linnaean hierarchy and its system of nomenclature. On the face of it, this may seem odd: biologists employ a system of classification whose theoretical basis has become obsolete. The continued use of the Linnaean system is more problematic when one considers that the system has lost its sorting rules, that cladistic revisions to the system render it less and less Linnaean, and that what remains of the system is flawed on pragmatic grounds. Given the current state of the Linnaean system and its importance in biological taxonomy, a philosophical investigation of that system is needed.

This chapter contains material that is reprinted, with kind permission from Kluwer Academic Publishers, from M. Ereshefsky, "The Evolution of the Linnaean Hierarchy," *Biology and Philosophy*, 1997; 12:493–519. Copyright 1997 by Kluwer Academic Publishers.

This chapter provides part of that analysis by sketching the evolution of the Linnaean hierarchy from Linnaeus's original system to the one used in contemporary biology. Section 6.1 provides an introduction to Linnaeus's original system, while Section 6.2 traces the changes that system has undergone from the rise of Darwinism to the formation of the Modern Synthesis. In more recent times, cladists have found fault with the Linnaean system. Section 6.3 reviews several changes offered by those cladists that want to preserve that system. Section 6.4 takes a quick look at some of the problems still facing the Linnaean system. The take-home message from this whirlwind tour will be that little of Linnaeus's original system is found in contemporary taxonomy, and much of what remains is problematic.

Given the current state of the Linnaean system one might wonder why biologists continue using it. Some speculations on that are offered in Section 6.5. The discussion in this chapter merely broaches the question of whether biologists should abandon the Linnaean system. The next two chapters directly address that question. Chapter 7 examines alternative systems of classification, and Chapter 8 asks whether it is reasonable to consider abandoning the Linnaean system given its entrenchment in biology. But before we can take up such matters we must start in the beginning, with Linnaeus himself.

6.1 LINNAEUS'S ORIGINAL SYSTEM

The importance of Linnaeus's work in the history of biological classification cannot be overestimated. Prior to Linnaeus, biological taxonomy consisted of conflicting practices: biologists disagreed on the basic categories of classification, how to assign taxa to those categories, and even how to name taxa. The result was a chaotic discipline marked by miscommunication and misunderstanding (Heywood 1985, 1; Mayr 1982, 173). Fortunately for biology, Linnaeus saw it as his divinely inspired mission to bring order to classification. The result is the Linnaean system of classification. That system offered clear and simple rules for constructing classifications, and it contained rules for naming taxa that greatly enhanced the ability of taxonomists to communicate.

The Linnaean hierarchy of twenty or more ranks found in contemporary biology is not the hierarchy of Linnaeus. The ranks of species and genus were commonly used before him by Cesalpino (Cain 1958,

158) and Tournefort (Atran 1990, 170). Linnaeus proposed expanding the number of types of biological taxa to five, offering a hierarchy of five categorical ranks for each kingdom. Those ranks are class, order, genus, species, and variety. As we shall see, Linnaeus had practical reasons for positing classes and orders: they help taxonomists remember the classification of genera and species. But Linnaeus had other, more philosophical reasons for positing a hierarchy of five levels, namely, he thought that nature generally consists of five fundamental ranks. He draws our attention to examples outside of biology (Larson 1971, 150). The discipline of political geography concerns kingdoms, cantons, provinces, territories, and districts; military science studies the actions of legions, cohorts, maniples, squads, and individual soldiers; and philosophy groups entities by *genus summum, intermedium, proximum, species*, and finally the individual.

Along with his hierarchy, Linnaeus provided a set of rules for sorting organisms into taxa. An important desideratum of scientific classification in the eighteenth century was the Aristotelian method of logical division (Cain 1958, 146ff.; Larson 1971, 146). The method of logical division was used by Linnaeus with the intention of revealing the true nature of organisms rather than their superficial resemblance. The method distinguishes five (yes, the number five again) types of predicates that can be attributed to a kind term: a definition, a genus, a differentia, a property, and an accident. The definition states which characteristics an entity must have to be a particular kind of entity. A definition, in other words, tells us the real essence of a kind's members. The genus is part of a kind term's definition but also applies to other kind terms. Using the stock Aristotelian example, the genus animal applies to both the species human and other types of animals. A differentia distinguishes one species in a genus from others in that genus. In the case of human, rationality distinguishes our species from all other species of animals. Putting the genus and differentia of human together, we get that species' definition, that of being a rational animal. Properties and accidents are, respectively, the necessary and accidental properties of essentialism (Section 1.1). Properties are characters found in all the members of a kind and stem from a kind's essence. Accidents have no necessary connection with a kind's essence. For example, the disposition for mathematical reasoning may be a necessary property of human, but wearing argyle socks is not.

Linnaeus used the method of logical division as the foundation for his method of classification. An organic species is distinguished by its differentia from the other species in its genus (1731[1938], sec. 257). Yet the "real distinction," or essence, of a species cannot be given without the definition of its genus (1731[1938], sec. 256). Thus we have the Aristotelian notion of a species' essence: its generic definition plus its species' differentia. Linnaeus also distinguishes those characteristics that are essential for membership in a species (whether they be part of that species' real essence or its necessary properties) from those that are accidental (1731[1938], sec. 258). The method of logical division is also used by Linnaeus to distinguish genera in an order, and orders in a class. Each order and each class has its own definition, and the subordinate taxa within them are marked by their distinctive differentiae (Cain 1958, 148; Larson 1971, 57ff., 73ff.).

The method of logical division is merely a logical schema for classification. To construct classifications, a theory concerning which traits serve as differentiae must be given. Linnaeus believed that the various characters of a plant's sexual organs – its fructification structure – best serve as differentiae for classification in botany. Here Linnaeus follows in the footsteps of Cesalpino and Tournefort (Cain 1958, 151–2; Larson 1971, 146; Atran 1990, 170). Both classified plants by fructification characters, and both offered reasons why one ought to do so. Linnaeus adopted those reasons and used them to justify his own sexual system. Following Cesalpino, Linnaeus believed that plants have two vital functions: the function of nutrition which preserves the individual, and the function of reproduction which preserves the kind. To know what kind a plant is one needs to study its function of reproduction, in particular, those parts that play a role in its reproduction. The other reason Linnaeus prefers classifying by fructification characters comes from Tournefort and is a practical one: fructification characters are easy to work with. Fructification structures are the most complex organ-system of plants and provide a large number of characters for study. Furthermore, such structures can be described with precision.

Linnaeus used thirty-one sexual characteristics (Atran 1990, 171). Flowers are divided into calyx, corolla, stamen, and pistil. Fruit contain three main parts: pericarp, seed, and receptacle. Each of these parts can be divided into further parts: seven for the calyx, two for the corolla, eight for the pericarp, three each for the stamens and pistils, and four each for the seed and receptacle. Linnaeus measured these thirty-one parts according to four variables: number, shape, proportion, and position. He calculated that these thirty-one parts and four variables would "suffice for 3,884 generic structures, or more than will ever exist" (*Philosophia Botanica* 1751, sec. 167; quoted in Atran 1990, 171). Linnaeus believed that these generic structures represented the total number of possible genera. Yet, as we shall see shortly, he was fairly certain that the number of existing genera was merely in the hundreds.

How did Linnaeus use these fructification characters for distinguishing taxa? Classes are distinguished by the numbers and positions of their stamens. Orders are separated by the numbers of their pistils (Larson 1971, 57–8). The differentiae of genera are a bit more complicated. The genus is the taxonomic level at which the primary function for maintaining biological kinds is fully displayed, namely, in the fructification or reproductive systems of organisms. Consequently, Linnaeus used all thirty-one characters to differentiate genera (Larson 1971, 74ff.). If genera are differentiated using all thirty-one fructification characteristics, what is left to distinguish species within genera? Here Linnaeus turns to the other vital function of plants – the function of nutrition. Linnaeus uses the parts involved in that function as well as more general characteristics of plants (for example, roots, stems, and leaves) to differentiate species (Larson 1971, 115ff.).

With the types of characters for identifying taxa in place, Linnaeus suggested a method for recognizing the essences of particular genera and species. The method starts with genera, following Linnaeus's belief that the fructification systems of genera are responsible for the existence of biological kinds. The first step is to select a familiar species of the genus under investigation and note the characteristics of its fructification system.[1] Next, one compares those characteristics with the fructification systems of other species in the genus. Discordant characters are rejected as possible essential characteristics of the genus. The more species examined, the more likely it is that the fructification system found is the essence of the genus. Not until every species in a genus has been examined can one know the essence of that genus.

The need to survey all the species of a genus has repercussions for knowing the essence of a species. A specific essence is composed of its genus's unique fructification system and whatever marks distinguish it from other species in that genus. So to know the essence of a species, one must know its genus's essence. But to know a genus's essence, the fructification system of every species within that genus must be surveyed. By parallel reasoning, though I have no evidence that Linnaeus asserted this, to know the differentia of a species one must survey all species within a genus. For as Linnaeus points out, the differentia of a species distinguishes that species from all others in its genus (1731[1938], secs. 257, 297).

The standards that Linnaeus placed on knowing the essences of genera or species are hard to meet. Linnaeus was well aware of this problem (Cain 1959a, 159–60; Larson 1971, 148ff.; Atran 1990, 174–5). He often lacked representatives of all the species of a genus and thus was unable to determine the unique fructification system of a genus. Given the method of logical division, a classification cannot be considered real or natural unless the true fructification systems of genera are determined. Consequently, Linnaeus saw his classifications as artificial and provisional guides for yet-to-be determined true classifications. Linnaeus thus found himself in a common occupational hazard of taxonomists: the disparity between the ideal standards one invokes and the actual methods one must employ in constructing classifications. Despite the disparity between theory and practice, Linnaeus's sexual system was a breakthrough. The fructification parts he recommended for classification were relatively easy to identify and describe. And unlike previous systems of classification, they allowed workers to arrive at similar conclusions. Given these pragmatic benefits, Linnaeus's system was adopted quickly (Mayr 1982, 178).

Thus far we have looked at Linnaeus's methods for determining taxa and their essential characters. Linnaeus also introduced a set of rules for naming taxa. Prior to Linnaeus, little agreement existed among biologists concerning the proper way to name taxa. As a result, taxonomists often gave different names to the same taxon. In addition, they often gave taxa unwieldy names that were hard to remember. Consider the following species name: "the *Dianthus* with stalks that have one flower, the scales of the clayx ovate and obtuse, the corolla many-cleft, and the leaves linear" (Larson 1971, 122). In an effort to provide unity and clarity to biological nomenclature, Linnaeus offered his own rules of nomenclature. I will discuss only his most famous one, the prescription that every species be given a binomial name.

All species names contain two parts: the first is the generic name, the second is the specific name. In the name of our species, for example, *Homo* is the generic name and *sapiens* is the specific name. Providing a species with a generic name was of the utmost importance for Linnaeus. He writes that "[a] specific name without a generic name is like a bell without a clapper" (1731[1938], sec. 286). Linnaeus offers several reasons that each species should have a generic name. The first stems from the metaphysical role of genera. Genera display the function of reproduction, and that function is responsible for the existence of genera and species. Second, Linnaeus thought that a species' name should indicate how that species varies from all other species. According to Linnaeus, that goal can be achieved only if the generic name of a species is included in that species' name (1731[1938], sec. 286). Consider the species *Rhododendron sikkimense*. The specific name *sikkimense* distinguishes that species from other species in the genus *Rhododendron*. But it does not mark a difference between that species and species in other genera. To do that we must write the generic name within a species' name.

Linnaeus's third reason for employing binomial names arises from his conviction that a biologist should memorize the taxonomic positions of all the species in the kingdom he studies (Cain 1958, 156ff.). Linnaeus realized that if species were just assigned specific names there would be too many species names to memorize. For example, he recognized approximately 10,000 species in the plant kingdom (Atran 1990, 171). Linnaeus recognized a much smaller number of genera: approximately 300 plant genera and 312 animal genera (Mayr 1969, 344). Moreover, he did not think those numbers would greatly increase: he held that the number of possible genera in the world was set once and for all in God's original creation (see discussion later in the chapter), and he believed that the world's major ecological zones had the same kinds of flora and fauna (Mayr 1982, 172). Given that a relatively small number of genera will exist, the placement of a generic name in a species' name serves as a handy guide for determining a species' classification. Once a biologist has memorized the placement of genera within a kingdom, she merely needs to read a species' binomial name to know its taxonomic position.

With Linnaeus's naming and sorting rules introduced, we can now turn to the theoretical assumptions that serve as the foundation of Linnaeus's system. Three theoretical assumptions stand out: creationism, essentialism, and the belief that genera are the most important taxa in his hierarchy. We have already discussed the status of genera in Linnaeus's system and we will return to the status of genera in later sections. Let us turn to Linnaeus's creation theories.

His earliest creation theory is found in the *Systema naturae* (1735).[2] According to Linnaeus, the entire world was initially covered with water except for an equatorial island. This island was exceptional in that it contained all of the habitats found on present day earth: alpine meadows, high plains, swamps, and so on. After creating the world, God set upon creating organisms to fill the habitats of that island. He created a single pair for each sexual species and a single hermaphrodite for each asexual species. After that initial act of creation, God created no new species. Linnaeus completes his creation theory by stating that in time the seas covering the world receded and the organisms on the island scattered throughout the world. One might wonder why Linnaeus thought that all of life was created on a single island rather than being distributed throughout the world. The answer, writes Linnaeus, is found in the Bible. The Bible tells us that Adam knew the names of all animals. Linnaeus reasoned that Adam could do that only if all of life were created in a single restricted area (Larson 1971, 96).

In the 1740s, Linnaeus came to realize that new species did arise after God's original creation (Larson 1971, 100ff.). This posed a problem for his creation theory. To address the origin of new species within the context of creationism Linnaeus offered another creation theory. He later codified this theory in the sixth edition of *Genera Plantarum*.[3] In the beginning, God created a pair of organisms for a single species in each genus. Over time, the members of those initial genera hybridized and gave rise to the existing diversity of species. Three aspects of this theory are worth highlighting. First, the hybridization that Linnaeus suggested was intergeneric; it was the mixing of genera that gave rise to new species. Second, because the number of genera was fixed in the beginning, so was the number of possible intergeneric hybridizations. In other words, once the original pair for a species in each genus was created, the number of species was established once and for all. Third, species for Linnaeus were preordained slots that may or may not have members. Some species had members in the beginning; however, many did not. Nevertheless, the nature of species lacking instances was fixed.

Later in life, Linnaeus altered his creation theory by limiting the creation of original organisms to representatives of even higher taxonomic categories (Larson 1971, 107ff.; Hull 1985, 111). In 1767, he suggested that God originally created a single pair for each order. Those pairs hybridized and gave rise to various genera; in turn, the members of those genera hybridized and gave rise to various species. Near his death, Linnaeus even suggested that God merely created a single representative pair for each class: "God created classes, from their mixture orders, from the orders genera, from the genera species" (from the *Skrifrer of Carl von Linne*, 2, p. 119 [1906]; quoted in Larson [1971, 111]).

The Evolution of the Linnaean Hierarchy

Although Linnaeus's views on creationism evolved, he never wavered on assumption that the number of species and genera, indeed all taxa, was fixed forever (Atran 1990, 179). This assumption underwrites Linnaeus's motivation for binomial names. Binomial names, recall, were introduced so that a biologist could memorize the positions of all the species within a kingdom. That goal is achieved only if the number of genera is sufficiently small that all of their names can be memorized. If the world contained a manageable number of genera and if God ceased creating any new ones, then binomial names would serve their intended purpose.

The final metaphysical assumption of Linnaeus's system that I would like to discuss is essentialism. Essentialism was the predominant approach to classification in the eighteenth century (Cain 1958, 146ff.; Larson 1971, 146). A central tenet of essentialism is that the world's constituents belong to natural kinds as the result of kind-specific essences (Section 1.1). Consequently, such essences reflect how the world is carved. From this tenet flows the methodological prescription that taxonomists should construct classifications by sorting entities into kinds according to their kind-specific essences. Linnaeus's own methods for sorting organisms into taxa are premised on these essentialist assumptions. The method of logical division provides the formal requirement that the real definition of a natural kind term refers to that kind's essence. Linnaeus's sexual system describes the empirical nature of those essences. As we have seen, the essence of a genus is its distinctive fructification system. The essence of a species is its genus's fructification system plus whatever features distinguish it from other species in its genus. Linnaeus also provided a list of accidental properties – properties that "do not make real distinctions" among kinds. They include such characteristics as size, color, scent, taste, and duration (1731[1938], sec. 121). In brief, the method of logical division and Linnaeus's sexual system are just extensions of essentialism.

Linnaeus's species and genera have many other features associated with essentialist kinds. One is that a natural kind may or may not have instances. The kind gold, for example, may or may not lack instances; it all depends on whether there are entities in the world with the essential atomic structure of gold. Linnaeus's later accounts of creationism held that many species lacked instances in the beginning, and once the appropriate hybridization occurred those species then had instances. Another feature of essentialist kinds is that the members of such kinds can come and go. That is, essentialist kinds may first have instances,

207

then lack them, and then have them again. Linnaean species have this feature as well. According to Linnaeus, the species *Veronica spuria* periodically lacked instances; yet whenever members of *Veronica maritima* and *Verbena officinalis* grew near one another and were subjected to appropriate circumstances, instances of *Veronica spuria* would arise again (Hull 1985, 49).

Another common essentialist tenet is that the number of natural kinds in the world is fixed once and for all. If you are a creationist you believe that number is set by God's original creation. If you are a contemporary astrophysicist you believe that it was set during the Big Bang. Linnaeus adopted the creationist version of this tenet and applied it to species: "All the species recognized by the botanists come forth from the Almighty Creator's hand, and the number of these is now and always will be exactly the same" (1731[1938], sec. 121). As we have seen, instances of Linnaean species can come and go, but the number and nature of those species is fixed forever.

6.2 DARWINISM AND THE MODERN SYNTHESIS

With Linnaeus's original system in hand, we can now turn to the evolution of that system since Linnaeus's time. This section sketches the changes that the system has undergone from the rise of Darwinism through the formation of the Modern Synthesis. The next section surveys some cladistic amendments.

As should come as no surprise, many of Linnaeus's metaphysical assumptions were shattered by the Darwinian revolution. In Linnaeus's earlier creationist theory, God created a pair of organisms for each species. In one of his later creationist theories, God created a pair for a single species in each genus, and the rest of the world's species are the result of intergeneric hybridization. With the advent of Darwinism and evolutionary theory we have very different accounts of speciation. For Darwin, species are not formed by "a special act of creation" (1859, 55) but are the result of evolutionary gradualism (Section 3.1). Natural selection causes some of the characters of a species to be replaced by others in a slow and gradual manner. When a sufficient number of appropriate changes has occurred, one species, or a portion thereof, evolves into another species. More recent accounts of speciation vary from Darwin's (Section 2.5). Mayr, Eldredge, and Gould, for example, believe that speciation is a much more rapid affair.

Mayr dubs them "genetic revolutions." Still, such accounts of specia-tion are premised on the process of natural selection, not divine creation.

With the rise of evolutionary theory also came the fall of essential-ism as a guiding principle to classification. Linnaeus believed that an organism belongs to a particular species because it has the appropri-ate species-specific essence. Since Darwin we have had a very differ-ent picture of species. For Darwin, the founders of the Modern Synthesis, and most cladists, species are genealogical entities. Mem-bership in a species turns first and foremost on an organism's genealog-ical connections to other organisms in that species rather than their qualitative similarity. The shift from essentialism to a genealogical per-spective primarily stems from the Darwinian idea that species are groups of organisms capable of evolving (Section 3.3). A species evolves only if its organisms are genealogically connected. Otherwise changes that occur in one generation of a species will not be transmit-ted to future generations and evolution will not take place. Genealog-ical connectedness is the linchpin of evolution. Similarity thus takes a back seat to relational properties in Darwinian classification. (See Chapter 3 for a full elaboration of this point.)

With the shift from essentialism and creationism to evolutionary theory comes a radically different concept of species. For Linnaeus, the number and types of species is forever determined by God's original creation. No new species can arise other than those preordained in God's original act. In Darwinian biology, the number and possible kinds of species that might evolve is open-ended. The organic world is viewed as one where natural selection and any number of fortuitous events (mutation, random drift, polyploidy, geological change) can cause speciation. There are no preordained slots in nature to be filled. This contrast between Linnaean and Darwinian species leads to another. For Linnaeus, if all the members of a species die, that species does not go extinct. A Linnaean species lacking members is merely an empty slot in nature to be filled when some appropriate intergeneric hybridization occurs. Darwin, on the other hand, writes that "[w]hen a species has once disappeared from the face of the earth, we have reason to believe that the same identical form never reappears" (1859, 131). In the *Origin*, Darwin takes this point to be an empirical one – the cir-cumstances that would cause the same species to exist again are extremely unlikely. Later in his life he treats this point as a metaphys-ical principle (Hull 1965, 50). On either reading, Darwin disagrees with

Linnaeus's belief that a species can lack members at one time and easily have members at a later time.

Another difference between Darwinian and Linnaean species is that the former but not the latter must be spatiotemporally continuous entities. Darwinian species are genealogical entities, that is, groups of organisms capable of evolving. Such evolution requires that the organisms of a species be spatiotemporally connected (Section 3.3). For Linnaeus, a species can be gappy: it can have instances at one time, no instances at a latter time, and instances once again at an even later time. Such species fail to form spatiotemporally continuous entities. This aspect of Linnaean species is a consequence of Linnaeus's belief in essentialism. Two organisms are members of the same species if they have the same species-specific essence, regardless of whether they are spatiotemporally connected.

One final ontological difference between Linnaean and post-Darwinian species is worth noting. In Linnaeus's system genera are the ontologically most important taxa. Hence the significant ontological divide in his hierarchy is between genera and other higher taxa. The importance of genera flows from two assumptions: fructification systems maintain the existence of biological kinds; and such systems occur at the level of genera. The importance of genera carries over to species. A species is distinguished from species in other genera by its genus's fructification system. Consequently, Linnaeus sees species and genera as real and natural entities because their existence depends on the fructification systems of genera (Cain 1958, 148, 152–3). Linnaeus was less sure of the existence of orders and classes. Classifications of such higher taxa, he suggested, were more artificial because their construction was in part determined by pragmatic considerations. For example, Linnaeus writes that "[a]n order is a subdivision of classes needed to avoid placing together more genera than the mind can follow" (*Philosophia Botanica*, sec. 161; quoted in Mayr 1982, 175).

The divide between genera and other higher taxa is no longer seen as the fundamental one in the Linnaean hierarchy. Its downfall stems from a couple of considerations. The first is the rejection of essentialism in Darwinian biology. Linnaeus thought that genera and species had real essential characteristics whereas classes and orders did not. Thus he put more stock in the existence of the former than the latter. In Darwinian biology no taxon of any rank has a taxon-specific essence. Species and genera lack essences just as much as classes and orders do.

So Linnaeus's first ground for distinguishing species and genera from other higher taxa no longer holds.

Second, an important distinction Linnaeus saw between species and genera versus classes and orders is that species and genera have generic-specific fructification systems. Those systems, Linnaeus thought, were responsible for the continued existence of all taxa. Classes and orders lack such systems and are merely aggregates of fructification systems that occur at the level of genera. Biologists no longer think that generic-level fructification systems are responsible for the existence of taxa. Instead many hold that the existence of taxa is the result of interbreeding among conspecific organisms (see, for example, Mayr 1970, 373–4, and Eldredge and Cracraft 1980, 89–90). Species are seen as populations whose members exchange genetic material through interbreeding. That interbreeding, it is argued, causes the populations of a species to evolve as a unit. In contrast, the species of a higher taxon do not exchange genetic material; the evolution of a higher taxon is merely the by-product of evolution occurring within its component species. So instead of Linnaeus's ontological divide between genera and other higher taxa, the authors of the Modern Synthesis, as well as many contemporary biologists, hold that the main ontological divide lies between species and all higher taxa. Species are actively evolving, cohesive entities – "*the* units of evolution." Higher taxa are aggregates of lower level processes – mere "historical entities."

The theoretical changes canvassed so far are important in themselves. They also have ramifications for Linnaeus's sorting rules, his rules of nomenclature, and the Linnaean hierarchy itself. For instance, the foundation of Linnaeus's sorting rules are his method of logical division and his sexual system. Neither method plays a role in contemporary biological taxonomy. Consider the method of logical division. It assumes that the members of species and genera have taxon-specific essences and that organisms should be divided into species and genera according to those essences. However, biological taxa are no longer seen as entities defined by qualitative essences. Instead, they are genealogical entities whose members share certain causal relations. If taxa lack essences, then the method of logical division has no role in classification – there are no essences on which it can operate. Accordingly, that method has been dropped from biological taxonomy.

Similar considerations apply to Linnaeus's sexual system. In Linnaeus's system, organisms are classified into a genus according to the qualitative similarity among their fructification systems. Genealogical connectedness is secondary and not a necessary condition for membership given Linnaeus's views on the recurrence of taxa. Since Darwin, the emphasis has been the opposite. Genealogical connectedness is considered primary for showing taxonomic membership; qualitative similarity between fructification systems, as any qualitative similarity, is secondary. Nevertheless, one might think that Linnaeus's approach to classification is not that distant from approaches since Darwin. Linnaeus's fructification systems are the reproductive organs of plants whereas the species of the Modern Synthesis are groups of organisms that can successfully interbreed. Aren't these approaches just two sides of the same coin? Not according to Mayr: "differences in the number of stamens and pistils, as practical as they may be for identification, are of little functional significance" in conspecific reproduction (1982, 178). Though such organs may facilitate pollination they do not serve as significant isolating mechanisms.

The theoretical shift in biological taxonomy since Linnaeus not only undermines his sorting rules but also his most important rule of nomenclature. Recall Linnaeus's motivation for the rule that each species be assigned a binomial. Linneaus thought that a biologist should memorize the taxonomic positions of all the species within the kingdom he studies (Cain 1958, 156ff.). Linnaeus realized that there were too many species to do that, but he thought that the number of genera in the world was sufficiently small. Furthermore, he thought that the number of genera in the world would remain constant because God had determined all possible genera in his initial act of creation.

Linnaeus did not envision a biological world where evolution continually gives rise to new species and genera. As a result, he drastically underestimated the diversity of flora and fauna in the world. Linnaeus held that a total of 312 genera and 4,000 species existed in the animal kingdom. More recent estimates cite well over 50,000 genera and 7 million species (Mayr 1969, 344) – obviously too many generic names to commit to memory. Mayr (1969, 344) observes that "a generic name no longer tells much to a zoologist except in a few popular groups of animals" (also see Hull 1966, 16). Linnaeus's assumptions concerning the diversity of flora are even further off target. He recognized approximately 6,000 species of plants and thought that the total number might approach 10,000 (Mayr 1982, 172). Contemporary botanists recognize hundreds of thousands of plant species; for example, over 200,000 species of phanerogams (flowering plants) are known to exist. Clearly

Kingdom

Phylum

Subphylum

Superclass

Class

Subclass

Infraclass

Cohort

Superorder

Order

Suborder

Infraorder

Superfamily

Family

Subfamily

Tribe

Subtribe

Genus

Subgenus

Species

Subspecies

Figure 6.1   The expanded Linnaean hierarchy.

Linnaeus's motivation for positing binomial names – that they allow a biologist to memorize the positions of all the species in a kingdom – has been lost.

Linnaeus's underestimation of the diversity of life affects his system of classification in other ways. Linnaeus posited five categorical ranks: kingdom, class, order, genus, and species. Biologists in the last hundred years have found that number of ranks insufficient for representing nature's diversity. Accordingly, they have added more ranks to the hierarchy. Simpson (1961, 17) and Mayr (1969, 91), for example, have expanded the Linnaean hierarchy to twenty-one ranks commonly used by zoologists today (Figure 6.1).

As we shall see in the next section, some biologists think that even these twenty-one ranks are insufficient for properly representing life.

Before leaving this section, let us review what is left of the original Linnaean system of classification. We have traced the evolution of the Linnaean system from the inception of Darwinism to the completion of the Modern Synthesis. During that time, three of its central onto-logical assumptions – creationism, essentialism, and the distinction between genera and other higher taxa – have been lost. Linnaeus's motivation for assigning species binomial names has become obsolete. Even his sexual system and the method of logical division have been abandoned. Of course, Linnaeus's original categorical ranks are still used, but biologists have supplemented those ranks with sixteen addi-tional ranks. By the end of the Modern Synthesis, all that remains of Linnaeus's original system is a hierarchy of categorical ranks and the use of binomial names. The evolution of the Linnaean system does not stop here. In the last thirty years cladists have argued that the Linnaean system is in need of even further revision.

6.3   CLADISTIC CHANGES

Cladists see a host of problems with the Linnaean system. Some of those problems stem from a conflict between cladism and the Linnaean system. Others are problems that affect the Linnaean system regard-less of whether one is a cladist or an evolutionary taxonomist. This section highlights various problems raised by cladists and illustrates how some cladists (Patterson and Rosen 1977, Wiley 1979, 1981, Eldredge and Cracraft 1980) have amended the Linnaean system in an attempt to resolve those problems. As we shall see, the Linnaean system continues to evolve. Cladistic revisions of the Linnaean system result in a system of classification that is even further removed from Linnaeus's original system. But before getting to those revisions it should be noted that not all cladists want to amend the Linnaean hier-archy. Some (for example, Hennig 1969 [1981], Griffiths 1976, and de Queiroz and Gauthier 1992, 1994) believe that the best way to resolve the problems facing the Linnaean system is simply to abandon that system. We will take up their motivations for rejecting the Linnaean hierarchy, as well as their alternative approaches to nomenclature, in later chapters.

The first cladistic problem to consider is the proliferation of Linnaean ranks. Significant branching on the tree of life has occurred many hundreds of times. Cladists would like to represent the phylogenetic taxa that result from such branching, but the twenty-one ranks of the Linnaean hierarchy established by evolutionary taxonomists are insufficient for doing so (Patterson and Rosen 1977, 155–7; Eldredge and Cracraft 1980, 221–2). Consider two examples. Patterson and Rosen (1977, 155–7) suggest that to classify the phylogeny of a group of herring requires the introduction of nine new Linnaean categories. This in itself may not seem like much of an increase, but Patterson and Rosen's suggestion concerns only a small segment of the Linnaean hierarchy, namely, nine new categories between the ranks of family and superfamily. Similarly, Gauthier et al. (1988, 60) suggest that a phylogenetic classification of Amniota that preserves the rank of class for Reptilia is forced to rank Aves as a genus. However, the taxon Aves itself contains extensive branching and would require an explosion of new Linnaean ranks between genus and species.

Cladists, of course, could simply invent more Linnaean categories to accommodate phylogenetic classifications, but they are reluctant to do so. Their reason for not doing so turns on pragmatic considerations (Wiley 1981, 204). The continual addition of more and more categorical ranks would cause the Linnaean hierarchy – the very framework of classification – to be in a constant state of flux. This is no small matter given that all workers using Linnaean classifications, whether they be systematists, ecologists, or geneticists, would continually need to learn the latest version of the Linnaean hierarchy.

Alternatively, some cladists suggest ways of constructing Linnaean classifications that avoid the creation of more Linnaean ranks. One such method is the technique of "phyletic sequencing" (Nelson 1972; Wiley 1981, 206). Phyletic sequencing minimizes the number of ranks assigned to monophyletic taxa on an asymmetrical cladogram by giving those taxa the same Linnaean rank. Phyletic sequencing nevertheless accurately captures the phylogenetic relations of those taxa by listing their names in the order of their branching. Consider an example from Eldredge and Cracraft (1980, 226–7). Figure 6.2 contains the cladogram for some major groups of vertebrates. A classification using phyletic sequencing that reflects that cladogram is the following:

> Superclass Agnatha
> Superclass Actinopterygii
> Superclass Amphibia

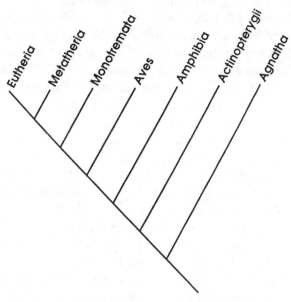

Figure 6.2   A cladogram of seven major groups of vertebrates. Adapted from Eldredge and Cracraft 1980, 226.

> Superclass Aves
> Superclass Monotremata
> Superclass Metatheria
> Superclass Eutheria

A classification in which phyletic sequencing is not employed requires five additional ranks:

> Superclass Agnatha
> Superclass Gnathostomata
>   Class Actinopterygii
>   Class Choanata
>   Subclass Amphibia
>   Subclass Reptilomorpha
>   Infraclass Aves
>   Infraclass Mammalia
>   Division Monotremata
>   Division Theria
>   Cohort Metatheria
>   Cohort Eutheria

Notice that in this traditional Linnaean classification there is not only a proliferation of ranks but also of taxa. Theria, for example, is added because it is the more inclusive taxon containing Metatheria and Eutheria. Hence another reason cladists advocate not representing all branching points with Linnaean categories: to do so requires the naming of both sister taxa at each branching point – for example, the naming of Monotremata and Theria.

Another problem cladists note with the Linnaean hierarchy is the need to revise previous classifications in light of new data, especially new data concerning fossil taxa (Wiley 1979, 323ff.; Eldredge and Cracraft 1980, 224). Wiley describes the problem this way:

> The very practical problem is that every time a fossil group is reassigned it demands a new subordination of ranks up and down the line (1979, 325).

> Even if most fossil groups were confidently assignable as their recent relatives . . . , it takes only a single fossil taxon of sufficiently ancient origin and controversial relationships to keep a classification in a constant state of flux (1979, 326).

This constant state of flux is in part caused by the Linnaean requirement that all taxa be assigned a categorical rank. The insertion of a newly discovered fossil taxon within a strictly Linnaean classification may require that its more inclusive taxa be bumped up a rank. If the newly discovered taxon is of a fairly low rank, that revision could be quite extensive. The actual revision of such classifications is an inconvenience in itself. And, as Wiley (1981, 217) notes, it is "simply too much to ask of colleagues" to relearn such classifications on a regular basis.

Some cladists suggest a way of avoiding such wholesale revisions. If those revisions are the result of assigning Linnaean categories to fossil taxa, then simply stop assigning fossil taxa Linnaean categories. Instead, merely insert such taxa in already existing classifications without a Linnaean rank. This is just the approach adopted by Patterson and Rosen (1977) and Wiley (1979; 1981, 219). They suggest that newly discovered fossil taxa be inserted in classifications right above their sister taxa. Furthermore, instead of assigning such taxa Linnaean ranks, those taxa should be given the non-Linnaean rank of "plesion." ("Plesion" is short for "plesiomorphic sister taxon.") The use of plesions certainly does the trick: more and more fossil taxa can be added to classifications without the need to revise the ranks of any of the other taxa in those classifications.

The above two modifications of the Linnaean hierarchy reveal a general approach employed by those cladists who want to preserve the Linnaean system: simply classify certain taxa in non-Linnaean ways. When newly discovered taxa are classified using phyletic sequencing, some inclusive taxa are neither classified nor given a Linnaean rank. When fossil taxa are assigned the rank of plesion, those taxa are not given a Linnaean rank. These changes are significant departures from the Linnaean system. First, they violate the Linnaean maxim that all taxa be given a Linnaean rank. Second, they cause cladists to construct classifications that are less and less Linnaean. As the number of plesions or sequenced taxa increases, the ratio of Linnaean to non-Linnaean taxa decreases. Indeed, when a cladist uses phyletic sequencing or employs the rank of plesion she is no longer actively constructing classifications within the Linnaean system: actual taxonomic revisions and taxonomic progress are conducted by purely non-Linnaean means. The strategy of *not* applying the Linnaean system to various aspects of the biological world is a common one cladists use in amending the Linnaean hierarchy.

Another problem with the Linnaean hierarchy stems from the desire to construct classifications that indicate ancestor-descendant relations. In particular, the Linnaean system lacks a means for designating a species as the stem species of a larger taxon. For example, we can see that the following Linnaean classification does not show that the ancestral (stem) species of family A is species E.

Family A
Genus B
Species D
Species E
Genus C
Species F
Species G

This was not a problem for Linnaeus because he did not attempt to indicate ancestor-descendant relations in his classifications. For him, taxa at different ranks are related by common aspects of fructification systems. The species of a genus, for example, share a similar fructification system, and the genera of a class have fructification systems containing a common aspect. It might be the case that hybridization between two species gives rise to a third species; yet Linnaeus did not attempt to capture such relations in his classifications.

Some cladists, particularly pattern cladists, argue that the desire to capture stem species in Linnaean classifications is merely a pseudo-problem. We rarely, if ever, know which species is the stem of a larger taxon. So we should not worry about the inability of Linnaean classifications to specify stem species. Wiley (1981, 223) disagrees. He believes that biogeographic information can help identify stem species. Moreover, he argues that "a general system of classification must be capable of handling *all* organisms and not just descendants." Accordingly, Wiley suggests that a stem species of a suprageneric taxon be placed in parentheses besides that taxon. Here is his example (ibid.).

> Supercohort Aves (*Avus ancestorcus*)
> Plesion Archaeornithes
> Cohort Neornithes
> Plesion Hesperonithiformes
> Subcohort Palaeognathae
> Subcohort Neonathae

Notice that of the seven taxa in this classification, three are classified by non-Linnaean means.[4]

There is one other cladistic amendment to the Linnaean hierarchy I would like to consider. Recall that for Linnaeus, taxa of a particular rank are comparable; that is, they have some common significant feature. The members of each genus, for example, display a common and unique fructification system. The members of each order share a common number of pistils. In contemporary biology, fructification systems, or parts thereof, are no longer taken as the sorts of traits that render the taxa of a single rank comparable (Section 6.2).

A number of alternative ways of making taxa of the same rank comparable have been suggested. Hennig (1965 [1994], 272) provides his own: taxa of the same Linnaean rank should originate in the same time period. Hennig draws an analogy with geology. Just as geological strata fall into different geological categories depending on their time of origin, biological taxa should be assigned Linnaean ranks according to their time of origin. Besides achieving comparability among taxa of the same rank, Hennig sees another virtue in his proposal. "It is a justifiable aim to perfect the phylogeny diagram by giving, not only the relative sequence of origin of the monophyletic groups, but also the actual time of their origin" (ibid.).

As many cladists note, the attempt to align Linnaean categories with absolute ages is a radical departure from Linnaeus's original notion of taxonomic rank (see, for example, Griffiths 1973, 338–9; 1974, 117–8; and Eldredge and Cracraft 1980, 222). For Linnaeus, all taxa of the same rank contain a similar diversity of less inclusive taxa; for instance, all classes have orders, genera, and species. This is not the case when taxa are ranked according to age of origin. Suppose classes are taxa that originate during the Late Cretaceous. Some of those classes may contain orders, genera, and species. Other taxa originating during that time period might be monotypic: consider those taxa that arose during Late Cretaceous that quickly went extinct before the occurrence of any extensive branching. So on Hennig's proposal, some classes would be monotypic taxa while others would contain an array of subtaxa.

Cladists are split on what to do at this point. Hennig (1966, 192) suggests that we should redefine the Linnaean ranks according to geological ages and abandon our intuition that taxa of the same rank should contain a similar degree of complexity. Most cladists, including later Hennig, shy away from this suggestion. They offer two further options. Some contend that biologists should simply stop using the Linnaean hierarchy (Hennig 1969; Griffiths 1974, 1976). We will get to this option later. Other cladists argue that we should continue using Linnaean ranks but drop Hennig's proposal of correlating ranks with geological ages (Eldredge and Cracraft 1980, 223; Wiley 1981, 214). These latter cladists suggest that we should no longer require higher taxa of the same rank to be comparable in any absolute sense. They recommend that higher Linnaean ranks merely indicate the relative age of taxa within a particular classification. For example, an order would be considered older than a family within the *same* classification, even though that order might be younger than a family in a *different* classification. Notice that this suggestion entails abandoning the Linnaean requirement that all taxa of the same rank be comparable. Thus in an effort to preserve the Linnaean hierarchy, some cladists choose to drop (or least scale back on) yet another tenet of Linnaeus's original system of classification.

So what is the state of the Linnaean hierarchy after the cladist amendments outlined in this section? Where possible, taxa should be classified by phyletic sequencing, and the classification of some more inclusive taxa should be ignored. Newly discovered fossil taxa should be given the non-Linnaean rank of plesion. Furthermore, the names of stem species should be written next to the higher taxon they give rise to rather than being listed in a subordinate fashion in the traditional Linnaean manner. Finally, Linnaean ranks no longer designate comparable taxa. For the most part, these changes to the Linnaean hierarchy are departures from Linnaeus's system rather than amendments to it: they prescribe methods of classification that are not Linnaean.

When we step back and consider these cladistic changes coupled with those of the Modern Synthesis, we get the following picture of the current Linnaean system. Pretty much all of its theoretical motivations are gone as the result of Darwinism. What remains is an amended hierarchy and system of naming taxa. According to cladists, however, that hierarchy cannot adequately handle various aspects of life (for example, its extensive branching, fossil taxa, and stem species). Consequently, some cladists classify those aspects of life with non-Linnaean means, though grafted onto more traditional Linnaean classifications. The only element of Linnaeus's original system that remains firmly intact is his binomial rule for naming species. But that rule, as we shall see, may need to be altered as well.

6.4   REMAINING PROBLEMS

Evolutionary taxonomists and cladists have drastically altered the Linnaean system to fit the demands of evolutionary biology and the tenets of their respective schools. One might think that biologists can now rest easy and employ the Linnaean system, even though it bears little resemblance to Linnaeus's original formulation. That belief is mistaken. The current Linnaean system, whether it be the version proposed by evolutionary taxonomists (Section 6.2) or that proposed by cladists (Section 6.3), still faces a number of pressing problems. This section presents some of those problems.

Naming Problems

Many of the remaining problems that face the Linnaean system stem from its rules of nomenclature. As we have seen, the rule of assigning species binomial names dates back to Linnaeus. That rule has remained untouched throughout the various alterations of the Linnaean system. In the last fifty years the Linnaean system has been given additional rules of nomenclature. Suffixes are now added to the names of various higher taxa to indicate their ranks. In the animal kingdom, for example, all names of tribes have the suffix-*ini* and all names of families have the suffix -*idae*.

Table 6.1. *A comparison of suffixes prescribed by zoological and botanical codes*

Rank	Botanical	Zoological
Class	-opsida (higher plants) -phyceae (algae) -mycetes (fungi)	None
Order	-ales	None
Superfamily	None	-oidea
Family	-aceae	-idae
Subfamily	-oideae	-inae
Tribe	-eae	-ini

The rules for attaching suffixes to the names of higher taxa are given in contemporary codes of nomenclature. Those codes are the *International Code of Zoological Nomenclature* (International Commission on Zoological Nomenclature 1985), the *International Code of Botanical Nomenclature* (International Botanical Congress 1988), and the *International Code of Nomenclature of Bacteria* (International Association of Microbiological Societies 1992). Each code provides its own list of suffixes for the names of higher taxa, and the suffixes they assign to taxa of the same categorical rank vary. Table 6.1 shows a comparison of some of the suffixes prescribed by the zoological and botanical codes (from Wiley 1981, 388).

The name of a higher taxon is formed by attaching the appropriate suffix to the name of the *type genus* of that taxon. A type genus, it should be noted, is not considered typical of the genera in a higher taxon in any qualitative sense. The type genus is that genus whose name is used to form the names of more inclusive taxa (Wiley 1981, 391). (More on the role of type taxa in Section 7.3.)

The Linnaean requirement of assigning binomials and rank-specific endings to the names of taxa causes a number of practical problems. Let us start with binomials. As we saw in Section 6.1, one of Linnaeus's motivations for introducing binomials was his belief that a biologist should memorize all the species within the kingdom he studies. Given the limited number of genera recognized by Linnaeus, and his belief that few genera were yet to be discovered, his motivation for requiring binomial names was reasonable. But as we saw in Section 6.2, Linnaeus drastically underestimated the diversity of flora and fauna in the world. There are far too many generic names and taxonomic positions for a botanist or a zoologist to memorize. Consequently, Linnaeus's original motivation for assigning binomial names has been lost. Nevertheless, the name of a species still must include a generic name.

One might grant that binomials no longer achieve their original function but wonder why that is a pressing problem. Could we not just treat the incorporation of a generic name in a species' name as an innocuous by-product of the Linnaean system? Unfortunately, Linnaeus's binomial rule is not so innocuous. To begin, it can place a biologist in the awkward position of having to assign a species to a genus before obtaining the proper empirical information for doing so. According to the binomial rule, a biologist must first determine the genus of a species before naming that species. But in some circumstances, biologists lack the appropriate information for making that assignment. A cladist, for example, may discover a species but lack sufficient evidence concerning whether that species is more closely related to a species in one genus or a species in another genus. Similarly, an evolutionary taxonomist may discover a species yet lack sufficient information concerning its ancestry and ecology to assign it to an appropriate genus. In both circumstances, the binomial rule forces biologists to assign species to genera even if insufficient evidence exists for making those assignments. Cain (1959b, 242) observes that "the necessity of putting a species into a genus before it can be named at all is responsible for the fact that a great deal of uncertainty is wholly cloaked and concealed in modern classifications." As an example Cain (1959b, 242) suggests that at least one quarter of all passerine bird species have been assigned to genera on the basis of inadequate evidence.

Another problem with binomials involves taxonomic revision. The activity of producing the correct taxonomy of the organic world includes revising previous classifications: species often must be reclassified among genera, genera reclassified among tribes, and so on. Taxonomic revision is the norm, not the exception, in biology. The binomial rule, as well as the rule of assigning rank-specific suffixes to the names of higher taxa, makes such revisions doubly hard. Not only must biologists change the taxonomic positions of taxa but also the names of taxa. Here are several examples.

According to the binomial rule, when a species is assigned to a different genus its generic name must be altered to reflect that species' genus. Thus, when the species *Cobitis heteroclita* was found to belong to the genus *Fundulus*, its name was changed to *Fundulus heteroclita* (Wiley 1981, 399). A more problematic case occurs when a species is assigned to a genus that already has a species with that specific name. In such cases, both a new generic name and a new specific name are needed. For example, the bee species *Paracolletes franki* was found to belong to the genus *Leioproctus*. Yet a species named *Leioproctus franki* already existed. So *Paracolletes franki* needed an entirely new name to avoid the existence of a homonym (Michener 1964, 183–4, 188).

One might agree that this is an irritating feature of the Linnaean system, but wonder about its extent. However, consider Michener's (ibid.) example of 288 species of bees being reclassified among three genera. Because the revision was performed according to the binomial rule, new names were assigned to all 288 species. More precisely, the generic name for each of the 288 species had to be rewritten, and 16 specific names had to be rewritten to avoid assigning the same binomial to different species. The number of names altered in this example is fairly large, and such cases of revision are far from isolated. Furthermore, as Mayr (1969, 344) notes, "[i]n these examples we are not dealing with the results of excessive splitting or any sort of arbitrariness, but with a serious weakness of the entire system of binomial nomenclature."

Taxonomic revisions occur at the level of higher taxa as well. In such cases, the rule of attaching rank-specific endings to the names of higher taxa makes such revision doubly hard. Once again, not only must those taxa be reclassified but their names must be altered. Workers in all major schools of taxonomy revise classifications by lumping and splitting higher taxa. Cladists are pressed to do so in an effort to eliminate paraphyletic taxa from earlier classifications (de Queiroz and Gauthier 1994, 27–8; Graybeal 1995, 247). Consider an example from de Queiroz and Gauthier (1994, 27–8). When a cladist is confronted with a classification in which the family Agamidae is paraphyletic (Figure 6.3a) she has two choices. She can lump all three crown taxa into a single family named "Chamaeleonidae" (Figure 6.3b). But then the taxon previously called "Chamaeleonidae" needs a new name to reflect its status as a less inclusive taxon. Alternatively, she can split the three crown taxa into separate families (Figure 6.3c). But then the names of the two crown taxa of the former paraphyletic taxon must be altered to reflect their status as families.

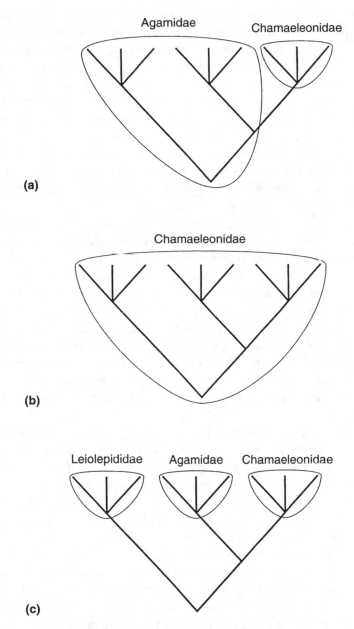

Figure 6.3 The reclassification of taxa to avoid the positing of a paraphyletic taxon. (a) The taxon Agamidae is paraphyletic and needs to eliminated. (b) One option is to lump the three crown taxa into the more inclusive taxon Chamaeleonidae. (c) Another option is to split Agamidae into two separate crown taxa: Leiolepididae and Agamidae. Adapted from de Queiroz and Gauthier 1994, 28.

As these examples illustrate, the Linnaean rules of nomenclature are themselves a source of instability in taxonomic revision: they require that the names of taxa be altered when taxa are reclassified. Biologists place a high value on taxonomic stability, yet the Linnaean system undermines such stability. The Linnaean rules for altering a taxon's name during revision may have further negative consequences: those rules serve as a disincentive to providing accurate classifications of nature. According to Michener, "most authors hesitate to change a clas sification in such a way as to require wholesale making of new combinations or new names" (1964, 186). Nevertheless, if they want to represent the organic world using the Linnaean system, such wholesale renaming must be done.

Taxonomic revision is not the only area in which the Linnaean rules of nomenclature render the activity of biological classification harder than necessary. Another problem occurs when biologists disagree over the rank of a single taxon. Such disagreements occur on various levels of the Linnaean hierarchy. One biologist may consider a taxon a species while another classifies it as a genus. At a higher level, a biologist may consider a taxon a family while another classifies it as a tribe. Such disagreements can result when biologists use different species concepts or subscribe to opposing schools of classification. Alternatively, different classifications can arise when workers subscribe to the same school of classification but use different techniques or employ varying data sets.

Consider a case where two biologists study a large group of organisms in a particular region. They both acknowledge that the members of the group successfully interbreed, produce fertile offspring, and are reproductively isolated from organisms in other such groups. Furthermore, they agree that the group forms a paraphyletic taxon. (See Section 4.1 for examples of such cases.) Although these biologists agree on the empirical details of the situation, they disagree on what rank to assign the group. One biologist adopts Mayr's (1969) Biological Species Concept and ranks the entire group as a single species. The other adopts Mishler and Donoghue's (1982) Phylogenetic Species Concept and treats the entire group as a genus consisting of two species. Because these biologists disagree on the rank of the group in question, the rules of Linnaean nomenclature force them to assign different names to what they agree is the same group. One biologist gives it a binomial name; the other assigns it a generic name.

The same problem arises when biologists disagree over genera. As

Michener (1964, 184–6) notes, South American bee specialists tend to be splitters: they recognize many small genera. Australian bee specialists tend to be lumpers: they prefer to classify bees into a small number of large genera. Because they disagree on which genera exist, they often disagree on the proper placement of a species in a genus: a lumper will place many species in one genus whereas a splitter will place those species in various genera. The binomial rule for naming species makes this situation even more confusing, for when lumpers and splitters disagree on the placement of a species in a genus they must assign different binomials to what they agree is the same species.

The Linnaean rule of assigning rank-specific suffixes causes even more confusing cases. Simpson (1963, 29–30) and Wiley (1981, 238) agree that the genus *Homo* belongs to a particular taxon, X. They disagree, however, on X's rank. Acting in accord with the Linnaean system, they attach different suffixes to the root *Homini* and give X different names: Wiley calls it "Hominini" and Simpson calls it "Hominidae." Their disagreement does not stop there. Wiley believes that X is a part of a more inclusive taxon, Y, which is a family. Using the root *Homini*, and following the rules of the Linnaean system, he gives taxon Y the name "Hominidae." So for Simpson and Wiley, the name "Hominidae" refers to two different taxa: for Simpson it refers to taxon X, for Wiley it refers to taxon Y. In brief, the Linnaean system causes Wiley and Simpson to assign different names to what they agree is the same taxon, and it causes them to give the same name to what they agree are different taxa.

Let us step back and summarize the problems stemming from the Linnaean rules of nomenclature. In Linnaeus's time, binomials served as guides for memorizing the placement of all the species in a kingdom. Binomials no longer serve that purpose. Instead biologists are saddled with several undesirable consequences of Linnaean nomenclature. First, the binomial rule requires that a biologist assign a species to a genus even though she may lack the proper information for doing so. Second, the binomial rule, as well as the rule that higher taxa be given rank-specific endings, makes taxonomic revision doubly hard: not only must the position of taxon be altered but so must its name. Third, the rules of nomenclature bring semantic confusion to taxonomic disagreements: single taxa are given multiple names and multiple taxa are given a single name. In brief, the Linnaean system makes the activity of classification more awkward and cumbersome than necessary. One might be tempted to dismiss these problems as merely pragmatic. But as we shall see in Section 6.5 and in Chapter 8, such considerations are paramount when biologists choose among systems of nomenclature, and rightly so.

Ontological Problems

Two more problems with the Linnaean system should be addressed. These problems indicate two ways that the Linnaean system may fail to capture true divisions in nature. The first is the assumption that life is hierarchically arranged. Recall from Chapter 1 that one requirement of hierarchical classification is that taxa (or kinds) of the same rank cannot overlap. That is, no two taxa of the same rank can share members. The type of species pluralism promoted in this book, and elsewhere (Kitcher 1984a, Ereshefsky 1992b, Dupré 1993), violates that requirement. According to the brand of pluralism advocated here, an organism may belong to two different species if those taxa are different *types* of species (see Section 4.1). For example, a single organism might belong to both an interbreeding species and a phylogenetic species, yet membership in those species is not identical. Consider the case of an interbreeding species that is paraphyletic. A phylogenetic approach to species either includes that interbreeding species within a more inclusive phylogenetic species or breaks that interbreeding species into less inclusive phylogenetic species (see Section 4.1 for examples of both situations). Either way, an organism can end up belonging to two species that do not completely overlap.

There may be further, less controversial grounds for doubting that all of life is hierarchically arranged. Michener (1963, 155) and Hull (1989[1964], 139–40) raise the issue of hybrid taxa. Members of different species from different genera have successfully hybridized and produced new species. The question, then, is to what genus does such a species belong? One possibility, according to Michener, is that a hybrid species belongs to both genera of its parental species. But that possibility violates the Linnaean requirement that an organism or a taxon cannot belong to two taxa of the same rank. Alternatively, Wiley (1981, 225–6) recommends placing a hybrid species in a new genus, or placing a hybrid species and its parental species in single genus. Wiley's recommendation would help preserve the requirement of nonoverlapping taxa. However, Michener (ibid.) responds that such a recommendation is motivated solely by the desire to maintain hierarchical classifications rather than any empirical evidence. Putting aside the question of what conventions one should adopt for classifying hybrid taxa, it should be recognized that hybrid taxa and their ancestral taxa are not hierarchically related. Hybrid taxa are the result of ancestral taxa coming together rather than branching. So the biological world, if only occasionally, violates the Linnaean assumption that nature is hierarchically arranged.

This brings us to one last problem. From Linnaeus's day to the Modern Synthesis and the inception of cladism, the ranks of the Linnaean hierarchy were intended to be more than just a set of hierarchically arranged words. Linnaeus thought that the major ontological divide in his hierarchy lies between genera and higher taxa. Species and genera contained genus-specific essences that made them real, whereas higher taxa were considered artificial groups of organisms. As we have seen, the rise of evolutionary biology led to the downfall of essentialism and the significance of fructification systems. Consequently, Linnaeus's ontological divide between genera and other higher taxa was abandoned. In its place, the founders of the Modern Synthesis suggested that the significant ontological divide in the Linnaean hierarchy lies between species and higher taxa (Mayr 1963, Dobzhansky 1970). Species are actively evolving cohesive entities bound by the process of interbreeding. Higher taxa, on the other hand, are merely artifacts of evolution at the species level. So while species are real and the "units of evolution," higher taxa are merely aggregates and "historical entities."

The ontological divide between species and higher taxa is not merely held by evolutionary taxonomists but is also suggested by some cladists. Consider Hennig's distinction between tokogenetic relations and phylogenetic relations (1966, 29–32). The different temporal stages of an individual organism are connected by ontogenetic relationships (Figure 6.4). The organisms of a particular species are linked by tokogenetic relationships: each individual is the offspring of two organisms of different sexes. As a result, the members of a species are connected by a network of bushlike relations in which the genes of multiple organisms are mixed. The taxa within a more inclusive taxon, on the other hand, are connected by phylogenetic relations. For instance, two species of a single genus are connected by their common ancestry and not by any mixing of genetic material among their current members. So once again, we have an ontological divide within the Linnaean hierarchy: species are cohesive wholes bound by interbreeding, whereas higher taxa consist of less inclusive taxa that share a recent common ancestry.

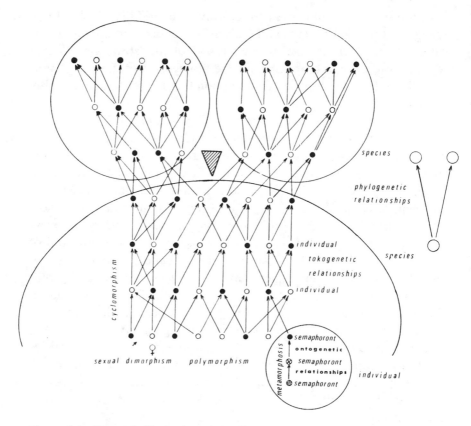

Figure 6.4   Hennig's illustration of the difference between ontogenetic, tokoge-netic, and phylogenetic relationships. The different temporal stages of an indi-vidual organism are connected by ontogenetic relations. The organisms in the different generations of a sexual species are linked by tokogenetic, or parent-offspring, relations. The species of a higher taxon are connected by phylogenetic, or ancestor-descendant, relations. From W. Hennig, *Phylogenetic Systematics*. Copyright 1966, 1979 by the Board of Trustees of the University of Illinois. Used with the permission of the University of Illinois Press.

As we saw in Chapters 3 and 4, controversy surrounds this onto-logical distinction. A number of biologists argue that many if not most species consist of populations whose members are not effectively con-nected by the process of interbreeding (for example, see Ehrlich and Raven 1969, Van Valen 1976[1992], Mishler and Donoghue 1982, and Templeton 1989). Their argument is encapsulated in two points. First, many species consist of asexual organisms that do not reproduce by interbreeding. Recall that such organisms are not the exception but the

prevalent type of organism on this planet, both currently and historically (Section 4.2). Second, many species of sexual organisms consist
of geographically separated populations that exchange little if any
genetic material through interbreeding. If these two points are correct,
then many species are not cohesive entities bound by the process of
interbreeding. In other words, many species are just like higher taxa
whose evolution is merely a by-product of processes that occur at lower
levels of biological organization. Thus, the ontological divide between
species and higher taxa may be going the way of Linnaeus's divide
between genera and other higher taxa. (See Ereshefsky 1991a for a
fuller discussion of this topic.)

The ramifications for the Linnaean system should be obvious. Since
Linnaeus, biologists have held that the ranks of Linnaean hierarchy
meant something more than just a set of words hierarchically arranged.
At the very least, the divide between the lower and higher categorical
ranks was supposed to reflect an ontological difference among taxa in
the world. But that has been cast in doubt. Linnaeus's distinction
between genera and other higher taxa no longer holds, and the more
recent evolutionary unit divide between species and higher taxa is
questionable. The history of biology, it seems, has left us with a hierarchy of categories without an ontological foundation.

The observation that the Linnaean categories correspond to little in
nature may seem radical, but it is far from new. In the *Origin*, Darwin
writes,

> we shall have to treat species in the same manner as those naturalists
> treat genera, who admit that genera are merely artificial combinations
> made for convenience (1859[1964], 485).

> I look at the term species as one arbitrarily given for the sake of con-
> venience to a set of individuals closely resembling each other, and that
> it does not essentially differ from the term variety (1859[1964], 52).

Ghiselin (1969, 93ff.) and Beatty (1985, 277ff.) interpret these remarks
as indicating that Darwin did not believe in the existence of the species
category (see Stamos 1996 for a dissenting view). More broadly, these
remarks suggest that Darwin doubted the existence of the Linnaean
categories. Of course, Darwin did believe in the existence of species
taxa, and he did maintain that such taxa could be named and classified
(Ghiselin ibid.; Beatty ibid.). He did not, for example, doubt the existence of the taxon *Homo sapiens*. But while Darwin was promoting the

existence of lineages and the process of evolution, he doubted the reality of the Linnaean categories.

6.5 PRAGMATIC CONSIDERATIONS

Let us take stock of the current state of the Linnaean system. As the previous sections have illustrated, almost every theoretical assumption of Linnaeus's original system has been abandoned. Creationism and essentialism have been replaced with evolutionary theory and historical approaches to classification. Linnaeus's sexual system and his use of the method of logical division have been replaced by the methods of phenetics, evolutionary taxonomy, and cladism. Even assumptions concerning the Linnaean ranks have experienced a considerable change. Linnaeus assumed that taxa of a particular rank are comparable; few biologists today hold that assumption. Linnaeus also assumed that the major ontological divide in nature resides between genera and other higher taxa; contemporary biologists place the divide elsewhere. The only theoretical assumption that continues from Linnaeus to contemporary biology is the assumption that all of life is hierarchically arranged. However, even that assumption may need to be tempered.

Despite the loss of so many of Linnaeus's theoretical assumptions, two key elements of his original system remain intact. For the most part, biologists use the ranks of species, genus, family, class, and order to indicate relationships of inclusiveness among taxa. And the vast majority of biologists still assign binomial names to species. These remaining components of Linnaeus's original system have been supplemented. Sixteen categorical ranks have been added to the Linnaean hierarchy, and rank-specific suffixes are now assigned to the names of many higher taxa. These additions to the Linnaean system are not so much a departure from Linnaeus's original system but supplements to it. Some of the changes suggested by cladists are a different matter. Those changes run counter to Linnaeus's original tenets. Phyletic sequencing and the placement of taxa under the rank of plesion classify taxa in non-Linnaean ways.

What is left of the Linnaean system is a shell of its former self. All that remains securely in place is a hierarchy of ranks and a set of naming rules. But as we saw in the previous section, those naming rules render the job of the taxonomist harder than necessary. In short, the

Linnaean system has lost its theoretical foundation, its rules of nomenclature are cumbersome, and various cladistic modifications of it are
non-Linnaean. Given the current state of the Linnaean system one may
wonder why the vast majority of biologists continue using that system
of classification. I cannot give a full answer to that question. Nevertheless, I will speculate on that answer by drawing on the work of some
historians and biologists.

A good place to start is with the initial acceptance of the Linnaean
system. The widespread acceptance of the Linnaean system in the late
eighteenth century has been attributed to the pragmatic virtues of that
system rather than its theoretical justifications (Larson 1971, Heywood
1985, Winsor 1985, Mayr 1982). Biological taxonomy was in a state of
chaos prior to Linnaeus. Biologists disagreed on how to construct classifications and even on how to name taxa. Consequently, different taxonomic studies employed noncomparable modes of representing taxa.
What made Linnaeus's system more than just another approach was
the clarity and simplicity of his rules. Consider Linnaeus's sexual
system for sorting organisms into taxa. As we saw in Section 6.1, they
are relatively easy to understand and to employ. The same goes for his
rules of nomenclature. Another boost for the Linnaean system was the
appeal of Linnaeus's revisions in botanical classification. In those revisions, Linnaeus unified conflicting botanical classifications, and he provided simpler names for approximately two-thirds of all known plants.
In sum, Linnaeus's rules and revisions brought order to a previously
disorganized discipline. That order, in turn, led to further benefits. It
laid the foundation for the "unprecedented flowering of taxonomic
research on animals and plants during the eighteenth and early nineteenth centuries" (Mayr 1982, 173).

Notice that the virtues just cited as responsible for the widespread
acceptance of the Linnaean system are pragmatic and not theoretic.
The popularity of Linnaeus's system did not stem from his adherence
to creationism and essentialism or his theory of hybridization, but from
the rules he gave for sorting and naming taxa. Further evidence for this
claim comes from that fact that Linnaeus's adherence to creationism
and essentialism were not a departure from the standard biological
theory of his day, whereas his rules for naming and sorting taxa were.
Of course, not all of those rules were original with Linnaeus (Section
6.1), yet the combination of them in a comprehensive and cohesive
system was.

Winsor (1985) makes a similar observation in her study of the continued use of the Linnaean system during the rise of Darwinism in the late nineteenth century. As we saw in Section 6.2, most of Linnaeus's theoretical assumptions were replaced by those of Darwin. Species are not the result of God's original creation but evolution by natural selection. Species are not immutable kinds but evolving lineages. Despite these and other theoretical changes, the Linnaean system maintained its hold in biology. According to Winsor, the Linnaean system persisted in the late nineteenth century because of "matters of practice" (1985, 84). Among those matters she highlights are Linnaeus's requirement of assigning only one name per taxon, "the simplicity and uniformity" of his publications, and the care Linnaeus took to provide thorough bibliographic references in his classifications.

When one turns to the reasons contemporary biologists give for adopting a particular system of nomenclature, they overwhelmingly cite pragmatic and not theoretical considerations. Consider the preamble of the *International Code of Zoological Nomenclature* (cited in Mayr 1969, 301):

> The object of the Code is to promote stability and universality in the scientific names of animals, and to ensure that each name is unique and distinct. All its provisions are subservient to these ends, and none restricts the freedom of taxonomic thought or action.

In other words, the code attempts to avoid taking sides in theoretical issues but instead tries to promote various pragmatic virtues in nomenclature. Stability is enhanced when name changing is avoided, and communication is promoted when biologists use one and only one name per taxon. The code adopts many of Linnaeus's naming rules for just these pragmatic reasons.

Contemporary advocates and detractors of the Linnaean system also highlight pragmatic reasons for adopting a system of nomenclature. Wiley (1979, 309) suggests that his Annotated Linnaean Hierarchy should be adopted because it is "designed to be maximally simple, informative, and stable." He makes no mention of the theoretical virtues of adopting his revised Linnaean system. The same sort of reasoning can be found in de Queiroz and Gauthier's (1992, 1994) arguments for abandoning the Linnaean system. Like Wiley, they highlight the practical values of a system of nomenclature:

> Taxonomy is fundamental to biology. It provides a reference system that permits communication and access to the literature, as well as a context

for comparative biology. In order to carry out these functions effectively, taxonomy must be governed by a unified body of concepts and principles designed to accomplish its practical goals within an appropriate theoretical context (de Queiroz and Gauthier 1994, 30; also see 1992, 472).

As de Queiroz and Gauthier suggest, some theoretical considerations should affect our choice of a nomenclatural system. Rules of nomenclature should not contradict the theoretical assumptions one uses when constructing a classification. We need to be careful here: a system of nomenclature and theory can conflict in two ways. The first is a conflict between the original theoretical assumptions behind a system of nomenclature and the theoretical assumptions a biologist makes when constructing classifications. For example, a Darwinian may construct classifications with the Linnaean system even though she rejects some of Linnaeus's original assumptions (species are viewed as evolving lineages rather than as static kinds, and so on). Nevertheless, the rules of nomenclature provide an adequate Darwinian classification because those roles are adequate in different theoretical contexts. A different type of conflict between a system of nomenclature and theory is less benign. Rules of nomenclature themselves can contradict the theoretical assumptions used by a taxonomist. For example, biologists might be in a bind if they were committed to a system of nomenclature that allows only hierarchical relations among taxa if much of life itself were not hierarchically arranged. Thus when Eldredge and Cracraft (1980, 166ff.) address the question of the usefulness of the Linnaean system in contemporary biology, they cite its practical values and that it allows us to represent the hierarchical arrangement of life.

Returning to our main question – Why do biologists keep using the Linnaean hierarchy despite the general abandonment of Linnaeus's theoretical assumptions? – we have a twofold answer. First and foremost, that system has a number of practical benefits. Second, Linnaeus's theoretical assumptions do not render his rules of nomenclature inconsistent with Darwinian and contemporary biology. The only aspect of Linnaeus's system that is anchored to a theoretical assumption is that taxa of different ranks are hierarchically arranged. For the most part, Linnaeus's system of classification is theoretically innocuous. Its continued use, from Linnaeus's time to the present, stems from its real and alleged practical benefits.

Still, the Linnaean system does have a number of pragmatic problems. As we saw in the previous section, those problems stem from the system's requirement that a taxon's name reflect its taxonomic position. Binomial names of species must contain generic names, and the names of most higher taxa must contain rank-specific suffixes. The problems that arise from these rules are pressing because they cause the Linnaean system to violate the primary aims of the *International Code of Zoological Nomenclature* as well as those practical virtues cited by contemporary promoters of the Linnaean system. Consider the preamble of the Zoological code quoted earlier: "The object of the Code is to promote stability and universality in the scientific names of animals, and to ensure that each name is unique and distinct." As we have seen, taxonomic revision is the norm and not the exception in biological classification. We would like classifications to remain as stable as possible, but such instability is an epistemological problem that cannot be eliminated from biological taxonomy. The instability of a taxon's name, on the other hand, can be avoided. Such instability is the result of the Linnaean requirement that a taxon's name contain information about its taxonomic position. If we separate the tasks of naming a taxon from that of indicating its position, such instability would not occur.[5]

Another aim cited in the preamble of the *International Code of Zoological Nomenclature* is that each taxon be given a distinct name. The Linnaean system violates this aim when biologists disagree on the placement or rank of a taxon. When they disagree on what genus to assign a species, they must, according to the Linnaean system, give that species different binomial names. The same problem occurs when biologists disagree over the rank of a higher taxon; in such cases they must attribute different names to a taxon. As noted in the last section, the use of the Linnaean system in such disagreements can also cause biologists to assign the same name to different taxa. These problems can be avoided and the distinctness of a taxon's name can be preserved if we drop the Linnaean requirement of incorporating the position of a taxon in its name.

In conclusion, the Linnaean system is a threefold system of theoretical assumptions, sorting rules, and rules of nomenclature. Over time, that system has lost its theoretical foundation as well as its sorting rules. Cladistic revisions have left it less and less Linnaean. And what remains of the system is flawed on pragmatic grounds. Taking all of these factors into account, it is high time to consider alternative systems of nomenclature. A handful of alternatives have been suggested in the last thirty years (for example, Michener 1963, 1964; Hull 1966; Hennig 1969[1981];

Griffiths 1974, 1976; Brothers 1983; de Queiroz and Gauthier 1992, 1994; and Ereshefsky 1994). Contemporary defenders of the Linnaean system are well aware of such alternative systems. They have responded that even though the Linnaean system has some practical disadvantages, alternative systems are no better and indeed worse (Mayr 1969, 345–6; Wiley 1979, 1981, Chapter 6; Eldredge and Cracraft 1980, Chapter 5). The jury is still out. Given the widespread use of the Linnaean system and the problems facing it, an in-depth comparison of the Linnaean system and its alternatives is sorely needed. The next two chapters are devoted to that task.

7

Post-Linnaean Taxonomy

Much of the last chapter was devoted to highlighting the problems facing the contemporary Linnaean system. Those problems fall into two camps: theoretical and pragmatic. On the theoretical side, the original assumptions underlying the Linnaean system – essentialism, creationism, and Linnaeus's sexual system – have been rejected. The Linnaean categories themselves correspond to little if anything in nature; and their number is insufficient for representing life's diversity. On the pragmatic side, the Linnaean rules of nomenclature are sources of instability and ambiguity. Furthermore, they saddle biologists with the task of assigning ontologically vacuous and misleading ranks to taxa.

Given the theoretical and pragmatic problems facing the Linnaean system, should biologists continue using that system? The problems outlined in the previous chapter are sufficiently pressing that we should at least study the feasibility of replacing the Linnaean system with an alternative. However, this is no small task, nor should it be done lightly. The Linnaean system is firmly entrenched in biology as well as popular culture. Critics of the Linnaean system must build a persuasive case for rejecting that system. They must develop viable alternatives and show that at least one of those alternatives faces fewer problems than the Linnaean system. But that is not enough. Given the pervasiveness of the Linnaean system, critics of that system need to provide more than theoretical reasons for adopting an alternative system. They need to show that switching to an alternative system is practically feasible.

The future of biological nomenclature is the focus of this and the next chapter. The current chapter provides an introduction and examination of alternative systems of nomenclature. As we shall see, various alternative systems, or better put, pieces of systems, have been suggested.

Section 7.1 discusses non-Linnaean methods for representing hierarchical relations. Section 7.2 examines alternative methods for naming taxa, and Section 7.3 discusses ways of defining taxon names. The aim of this chapter is to whittle those alternatives to a single, cohesive post-Linnaean system. The next and final chapter of the book compares that alternative system to Wiley's (1979, 1981) annotated Linnaean hierarchy. It also discusses what practical impediments stand in the way of switching from a Linnaean to a post-Linnaean system.

7.1 HIERARCHIES

A fundamental assumption of the Linnaean system is that life is hierarchically arranged: organisms belong to species, species to genera, genera to families, and so on. Although an organism belongs to multiple taxa, an organism cannot belong to two taxa of the same rank. Linnaeus's preference for hierarchical classifications stems from the a priori assumption that God created a world of hierarchically arranged organisms and taxa. To reveal that hierarchy, Linnaeus advocated using the Aristotelian method of logical division (Chapters 1 and 6). Taxa are distinguished by differentia, and an organism is sorted into taxa according to which differentae it instantiates, starting with its order, then its family, its genus, and finally its species. Contemporary biologists do not use the method of logical division. Nevertheless, they prefer hierarchical classifications as well. Their justification for that preference is the empirical claim that the organic world forms a genealogical hierarchy (for example, see Simpson 1961, 15, and Hennig 1966, 20). The contemporary motivations for hierarchical classifications differ from Linnaeus's in two important respects: they are posed as empirically testable claims, and they cite the genealogical nature of taxa as their justification.

The alternatives to the Linnaean system illustrated in this section do not challenge the assumption that life is hierarchically arranged. Instead, they challenge the assumption that the Linnaean categorical ranks should be used to capture that hierarchy. A number of reasons for scrapping the Linnaean categories were presented in the previous chapter. Here is a quick review of two of them.

As we have seen, the history of biology is littered with failed attempts to render the taxa of a single higher rank comparable. An order in one classification may occur at a very different level of inclusiveness from an order in another classification. A family in one classification may be much older than a family in another classification. The best we can expect is that the assignment of a higher Linnaean rank indicates the relative age or relative degree of inclusiveness of a taxon *within* a particular classification (Section 6.3). Higher Linnaean ranks, in other words, have no interclassificatory meaning; they are ontologically empty designations. Notice that the problem here is not merely ontological. The continued use of the Linnaean ranks misleads (at least some) biologists to think that taxa assigned the same rank are similar regardless of their classification. This mistaken view causes biologists to enter "unfruitful debates" over the ranks of taxa (Hennig 1969, xviii; also see Griffiths 1976, and Ax 1987, Chapter K).

Another reason for abandoning the Linnaean hierarchy is that the standard twenty-one ranks of that hierarchy are woefully insufficient for representing the complexity of life (Patterson and Rosen 1977, 155–7). Biologists could create more Linnaean ranks, but they are hesitant to do so because they would like to keep the framework of classification stable (Wiley 1981, 204). Annotated Linnaean systems (Wiley 1979, 1981; Eldredge and Cracraft 1980) avoid the need for more Linnaean ranks by *not* assigning Linnaean ranks to many taxa. Instead, such taxa are classified by phyletic sequencing and plesion ranking. However, as more taxa are classified with such techniques, classifications become less Linnaean. At best, the resultant classifications are an awkward mix of taxa classified by Linnaean methods and taxa classified by non-Linnaean means. At worst, classifications become dominated by taxa classified by non-Linnaean means. Either way, the result is a departure from the Linnaean system rather than a preservation of it. Instead of adopting such amendments to the traditional Linnaean system, it would be simpler and theoretically sounder to abandon the Linnaean ranks altogether.

Two alternative methods for representing the hierarchical relations are offered in the literature: indentation systems and numerical systems. We will start with the latter. One type of numerical system should immediately be placed to the side. Michener (1963) and Little (1964) suggest that taxon names be replaced with identification numbers, the hope being that such numbers would further the computerization of taxonomy. Although taxon names would become numerical, the use of Linnaean ranks would remain. This numerical system does not provide an alternative means for representing the hierarchical relations of taxa, so it will not be considered here. Furthermore, and this topic will be discussed in the next section, replacing the current names of taxa with new ones, numerical or otherwise, would be more of a hindrance than a help.

An alternative numerical system, suggested by Hull (1966), Hennig (1969[1981]), Griffiths (1976), and Lovtrup (1977), is more in line with current needs. In that system, the position of a taxon within a particular classification is indicated by a *positional number* rather a Linnaean rank. Consider Hennig's (1969[1981]) numerical classification of the Mecopteroidea. The phylogeny of seven taxa in that group is captured by Figure 7.1. The corresponding numerical classification is the following.

2.2.2.2.4.6	Mecopteroidea
2.2.2.2.4.6.1	Amphiesmenoptera
2.2.2.2.4.6.1.1	Trichoptera
2.2.2.2.4.6.1.2	Lepidoptera
2.2.2.2.4.6.2	Antilophora
2.2.2.2.4.6.2.1	Mecoptera
2.2.2.2.4.6.2.2	Diptera

Various types of information are captured by these positional numbers. The assumption that Mecoptera and Diptera are sister taxa is indicated by their numbers being the same except for their last digits. The hypothesis that Diptera and Antilophora are parts of the more inclusive taxon Mecopteroidea is captured by their positional numbers containing Mecopteroidea's number. And the assumption that Antilophora is a more inclusive taxon than Diptera is indicated by Antilophora's positional number containing fewer digits than Diptera's. The underlying rules governing positional numbers are these: sister taxa have the same positional numbers except for their last digits; subordinated taxa contain the positional numbers of their higher taxa; and more inclusive taxa have shorter positional numbers.

It is worth highlighting what positional numbers do *not* tell us. A taxon's positional number does not indicate that taxon's position in all classifications. Instead, a taxon's positional number merely represents that taxon's hierarchical position within a specific classification. Using de Queiroz's (1997, 132) example, Angiospermae might have the positional number 2 in a classification of anthophytes; yet in a classification of spermaophytes it is assigned the positional number 2.2.2.2.1. Positional numbers indicate a taxon's level of inclusiveness on a local level, not on a global level.

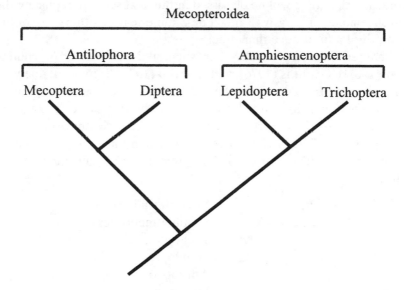

Figure 7.1 A hypothesis of phylogenetic relationships among mecopteroid insects. Adapted from Hennig 1969 (1981), 30.

This feature of numerical classifications gives them an advantage over their Linnaean counterparts. The Linnaean categories carry with them the misleading metaphysical assumption that taxa of the same rank are similar, that the taxa of a certain rank should have a common structure and a common degree of inclusiveness. A positional number, on the other hand, merely indicates the level of inclusiveness of a taxon *within* a classification. It does not carry the metaphysical connotations associated with the Linnaean ranks.

Another advantage of positional numbers is that they readily indicate the position of a taxon within a classification. The Linnaean system no longer does that. One of Linnaeus's original requirements was that the names of genera be committed to memory. If a biologist knows her genera, she can quickly determine the taxonomic position of a species. But as we saw in Section 6.2, there are far too many generic names to commit to memory. "Except for a small percentage of familiar taxonomic names, giving the name of a taxon conveys little or no information to anyone but a specialist in the group" (Hull 1966, 16). Positional numbers, on the other hand, allow a biologist from any field to locate the position of a taxon within a classification. A common refrain of systematists is that classifications should be repositories of taxonomic information that can easily be accessed by nontaxonomists. Numerical classifications better fulfill that aim than Linnaean classifications.

A further advantage of the numerical system is that it offers an internally consistent method for representing the complexity of life. As we have seen, the standard twenty-one ranks of the Linnaean hierarchy are insufficient for representing extensive phylogenetic branching. Annotated Linnaean systems capture that complexity by *not* assigning Linnaean ranks to many taxa. As a result, annotated Linnaean systems are inherently inconsistent: they "preserve" the Linnaean system by classifying taxa with non-Linnaean methods. Furthermore, they are merely stopgap measures because they only temporally preserve the Linnaean nature of classifications – by their own methods, classifications in the long run will contain mostly taxa classified by non-Linnaean methods. The numerical system is neither inconsistent nor a stopgap measure. Taxa are consistently classified with the aid of positional numbers, and that consistency will remain as taxonomic research continues.

Finally, positional numbers are more accommodating to taxonomic pluralism than Linnaean ranks. Consider the case of species. In Chapter 4 we saw that some taxa – paraphyletic interbreeding populations, in particular – count as species in evolutionary classifications but not in strictly phylogenetic classifications. On the Linnaean approach, a taxon cannot be a species in one classification but not in another. Here the numerical approach helps. Positional numbers merely indicate that a taxon is a basal taxon in one classification without implying that it is a basal taxon in another classification. Positional numbers merely indicate the level of inclusiveness of a taxon within a particular classification, not on a global level. If one is a supporter of taxonomic pluralism, then here is one more reason for preferring positional numbers over Linnaean ranks.

Of course, not all biologists are happy with the numerical system. Several supporters of the Linnaean system argue that the numerical alternative is inferior because "numerical prefixes are not the languages of ordinary use and are foreign to our efforts of communication" (Wiley 1981, 202; also Eldredge and Cracraft 1980, 224–5). I find it odd that a scientist would refer to numerals as "foreign" and not "ordinary." After all, numerals are one of the few forms of representation that the vast majority of scientists on this planet understand. At any rate, if numerals are foreign they are foreign because numerical prefixes are not commonly assigned to taxa. Such foreignness should not constitute grounds for rejecting a system of nomenclature. If it did, then scientific progress would grind to a halt because all new theories and modes of classification would be rejected because they are foreign.

Other objections to the numerical system have more merit. Some suggest that the numerical prefixes of lower taxa in large classifications may become inordinately long (Wiley 1981, 202; Eldredge and Cracraft 1980, 224–5; Ax 1987, 202). Imagine classifications containing positional numbers with fifty or more digits. This objection, however, turns on the false assumption that there will be "as many prefixes as taxon names" (Wiley 1981, 202). Numerical prefixes are assigned to taxa within particular classifications to indicate hierarchical relations *within* those classifications. The same positional numbers may be assigned to many taxa, albeit in different classifications. So the number of prefixes will be much less than the number of all named taxa.

Another objection is that the numerical approach requires the reassignment of positional numbers when taxonomic revision occurs, thus making revision doubly hard (Wiley 1981, 202–3; Ax 1987, 243). In other words, not only must taxon names be moved but the positional numbers associated with those taxa must be altered as well. This is certainly the case, but a similar if not worse problem affects Linnaean classifications. When a Linnaean classification is revised, taxon names are moved, taxa are assigned different ranks, *and* the rank-specific endings of taxon names (or generic names of taxa) must be altered (Section 6.4). Revisions are a twofold effort for the numerical system; they are a threefold effort for the Linnaean system.

Ax (1987, Chapter K) sees a better way to handle revisions, better than the methods offered in either the Linnaean system or the numerical system. He suggests using purely indented classifications. Farris (1976) introduced indented classifications using Linnaean ranks. Ax (1987) and Gauthier et al. (1988) adopt the use of indentations but drop the employment of Linnaean ranks. In purely indented classifications, relations of inclusiveness are indicated by indenting subordinate taxa below their higher taxa. Sister taxa have the same indentations. Here is a pure indentation classification of the group illustrated in Figure 7.1.

Mecopteroidea
  Amphiesmenoptera
    Trichoptera

Lepidoptera
Antilophora
Mecoptera
Diptera

The above spacing indicates that Mecoptera is subordinate to Antilphora, and that Mecoptera and Diptera are sister taxa. (For additional examples of pure indentation classifications, see Gauthier et al. 1988, and de Queiroz and Gauthier 1992; for an example of a hybrid indented/Linnaean system, see Farris 1976).

Like the numerical system, the indentation system has advantages over the Linnaean hierarchy. Indented classifications represent the hierarchical relations of taxa without the use of ontologically suspect Linnaean categories. Indented classifications free biologists from the need to create more Linnaean ranks. And the revision of such classifications can be performed merely by moving the names of taxa – there are no categorical ranks to be altered.

This last advantage, according to Ax (1987, 243), makes purely indented classifications preferable to numerical ones as well. A revised numerical classification requires the reassignment of positional numbers to already classified taxa. Ax suggests that we can avoid this problem by dropping positional numbers and replacing them with appropriate indentations. This suggestion, however, does not make indentation classifications preferable to numerical ones. When an indentation classification is revised, the indentations of the names of previously classified taxa must be readjusted. In other words, the vertical positions of taxon names as well as their horizontal indentations must be revised. Thus both numerical and indentation classifications require a twofold effort in revision.

Still, a proponent of the pure indentation systems might respond that more work is involved in resetting the positional numbers of taxa than just adjusting their indentations: there is more to write and keep track of when using positional numbers. But that extra bookkeeping – the assigning and reassigning of positional numbers – may make numerical classifications easier to use in some situations. A standard objection to the indentation system highlights the difficulty of comparing the hierarchical positions of taxa when their names appear on different pages of an extended classification. Merely looking at their indentations is insufficient; a ruler, Wiley (1981, 203–4) suggests, is needed to compare their positions. Proponents of the indentation

245

system are well aware of this objection (see Ax 1987, 244, and Gauthier et al. 1988, 60). They respond that such inconveniences are overshadowed by the general problems of the Linnaean hierarchy. Note, as well, that if multipage indented classifications are too cumbersome, then the Linnaean system is not the only fallback system. The positional numbers of the numerical system also provide a precise way to compare the positions of taxa within a classification.

We now have two non-Linnaean methods for representing hierarchical relations on the table. Neither is perfect, but then no system of nomenclature is. Nevertheless, both are an improvement over the Linnaean system. Numerical and indentation classifications adequately represent the hierarchical relations of taxa without using ontologically vacuous and misleading categories. They represent the complexity of nature without invoking an awkward mixture of Linnaean-ranked and non-Linnaean-ranked taxa. And they make taxonomic revision easier by avoiding the needless tasks of adjusting the categorical ranks of taxa and altering their names.

Let us suppose that a non-Linnaean system should be adopted. Which of the two alternative systems discussed here is preferable? Both the numerical and the indentation systems have the general advantage of dropping the Linnaean categories. But they have their drawbacks as well. The positional numbers of the numerical system must be altered during taxonomic revision. That disadvantage of the numerical system (and the corresponding one in the Linnaean system, namely, the alteration of ranks) can be avoided by employing indented classifications. Indented classifications, on the other hand, make it hard to compare the positions of taxa in multipage classifications. That problem can be avoided by using the numerical system when a classification spans more than a page. Perhaps, then, we should use the indented system for one-page classifications and the numerical system for longer classifications.

In the end, it does not greatly matter which system is used, so long as one or the other is used rather than the Linnaean hierarchy. Both the numerical and indentation systems overcome the major problems of the Linnaean hierarchy, and the individual problems of each are slight by comparison. Perhaps some biologists will use only numerical classifications, others will use only indented classifications, and still others will use one or the other depending on the situation. That's fine. What is important is that taxonomists construct classifications free of the metaphysically misleading Linnaean ranks and free of the pragmatic problems of the Linnaean system. Both the numerical and the indentation systems do that.

One aim of this chapter is to sift through various alternative systems of nomenclature and derive a single coherent post-Linnaean system. In an effort to do that, this and the following sections will conclude with recommendations for a post-Linnaean system of classification. Those recommendations will summarize the discussions of particular sections, and, in total, they will serve as an outline of the post-Linnaean system offered in this chapter. The resulting series of recommendations will be used as a template in Chapter 8 for comparing the post-Linnaean system with the contemporary Linnaean system.

Recommendations

R1. Taxa are no longer assigned Linnaean ranks.
R2. The hierarchical relations of taxa within a classification are indicated by either indentations, positional numbers, or both.

7.2   NAMING TAXA

The Linnaean ranks are just one area of the Linnaean system worth replacing. Another is the Linnaean rules governing the naming of taxa. A common theme running through those rules is that a taxon's name should indicate a taxon's classification (Section 6.4). Species are assigned binomial names that indicate their rank and genus. Many higher taxa are assigned suffixes that indicate their rank. These suffixes vary among kingdoms and in some instances highlight the kingdom of a taxon. Finally, the placement of a higher taxon is often indicated by incorporating the name of a type genus within its name.

A considerable amount of time was spent in Chapter 6 surveying the problems that arise from the custom of highlighting the rank and placement of a taxon with its name. For one, if the assignment of a Linnaean rank to a taxon does not reflect anything in nature, then indicating the rank of a taxon with its name is an empty and misleading gesture. Then there is an array of practical problems. When taxonomic revision occurs, not only must the placement of a taxon be altered but so must its name. And when taxonomists disagree on the rank of a single taxon, they are required to assign different names to what they agree is the same taxon. All of these problems arise from the general Linnaean requirement that the name of a taxon perform two functions:

naming a taxon and indicating that taxon's placement in a classification. Separate these two functions and the problems dissipate. More specifically, a taxon name should be just that; it should not be used as a device for indicating a taxon's rank or placement. Given this suggestion, what would be a suitable alternative method for naming taxa?

First, note that if one adopts either the indentation or numerical system, there is no need to incorporate the rank of a taxon within its name. Its hierarchical placement within a classification is indicated by either indentation or a numerical prefix. Thus the indentation and numerical systems nicely divorce the function of naming a taxon from classifying it. But we are still left with the question of how to name taxa in non-Linnaean systems. A handful of proposals can be found in the literature.

One suggestion offered in the 1960s is that the Linnaean names of taxa be replaced with numerical names (Michener 1963, Little 1964, Hull 1966). Hull calls such names "identification numbers." Using Michener's (1963, 166) example, "0325761 could represent *Musca domestica* and 0325762, *Felis leo*." Identification numbers provide no reference to the taxonomic placement or rank of a taxon, so they avoid the problems that arise when such information is incorporated in a taxon's name. Nevertheless, one might wonder whether the problems cited above can be avoided by some other means than the wholesale replacement of traditional names. Couldn't we just keep traditional names but simply ignore those features of taxon names that indicate taxonomic rank and placement?

As we shall see shortly, we could and should do just that. But there are other reasons given for adopting identification numbers. One is that biological taxonomy will come to rely heavily on computers.

> The exponential increase in the outpouring of systematic literature will probably have to go on for another decade or two until the problem of keeping track of it becomes almost completely hopeless before a suitable international center for storage and retrieval of systematic and related data is established. It is obvious that information-storage and retrieval systems will be enormously useful to all kinds of systematists and will probably use some sort of numbering system (Michener 1963, 166).

Michener correctly predicts that taxonomists will rely on computers in their work, but he offers the wrong reason. Computers have proved invaluable in constructing classifications from complex sets of data. Computations that would take months to perform by hand can be accomplished in hours. However, Michener's emphasis is on using computers as repositories of classifications. That use has not materialized. Consequently, the need for identification numbers has not materialized either.

But let us assume otherwise. Suppose computers are required for "the storage and retrieval of systematic and related data." Do we then need to replace traditional taxon names with identification numbers? No. Computers can store letters and names as much as they can store numbers. Perhaps that was not so obvious in the 1960s when FORTRAN, a number-based computer language, was a standard for research. At any rate, the current use of computers does not force us to replace traditional names with identification numbers. More important, such a replacement should be avoided because of the instability and inconvenience it would cause. If every taxon were given a new name, not only would biologists need to learn those names, every textbook and environmental law that mentions a taxon would need to be rewritten. This is not an instance of scientific change in which a few key concepts are replaced, as in the case of abandoning the Linnaean ranks. Here we are talking about casting aside a large chunk of the language of biology.

A more sensible proposal is to leave the names of taxa as they are, yet deny that those names indicate taxonomic position. Consider the suffixes and roots in the names of higher taxa. Griffiths (1976) suggests keeping the traditional names of higher taxa, yet abandoning the requirement that suffixes and roots indicate the ranks and positions of taxa (also see de Queiroz and Gauthier 1992, and Ereshefsky 1994). For example, the taxon Tipulidae would retain its name, but its suffix, -idae, would no longer indicate its rank; similarly the root in its name, Tipul, would no longer indicate that it contains the genus Tipula. If a taxon's name no longer indicates the rank and position of its taxon, then the problems encountered during revision and disagreement can be avoided. Tipulidae can be reassigned to a more inclusive position in a classification without the need to change its name. Furthermore, two biologists can disagree on its position in a classification without being required to assign it different names.

Griffiths's suggestion promotes stability in classification by keeping the name of a taxon constant before and after revision, and it avoids synonymy due to taxonomic disagreement. Furthermore, his suggestion promotes stability in taxonomy by retaining the traditional names of higher taxa even though it rejects the Linnaean categories and associated rules of nomenclature. Griffiths's proposal ensures progress in biological nomenclature but at not too high a price, as in the case of identification numbers.

We have considered post-Linnaean treatments of preexisting names of higher taxa. We should also consider post-Linnaean approaches to naming newly discovered higher taxa. Griffiths does not address this issue (but he does talk about the naming of newly discovered species – we will get to that issue shortly). The application of suffixes to the names of higher taxa should be considered optional. Taxa can be named without suffixes, or some biologists may consider it easier to produce new names by taking a common root for a series of taxa and applying different suffixes. Nothing greatly hinges on which approach a biologist takes, though one important cautionary note is needed. If a biologist chooses to use a single root with multiple suffixes to generate names, then that root and those suffixes are simply parts of names; they do not indicate the placement or inclusiveness of taxa.

Thus far we have considered the names of taxa above the Linnaean rank of genus, and our attention has focused on the use of suffixes in those names. Recall that four devices are employed by contemporary nomenclature for indicating the rank of taxa: using suffixes for the names of higher taxa above the level of genus; italicizing the names of genera; assigning binomials to species; and assigning trinomials to subspecies. Having dispensed with the assumption that suffixes indicate taxonomic rank, let us turn to the other devices employed by traditional nomenclature.[1]

If the Linnaean rank of genus is dropped in post-Linnaean classification, then no taxa should be distinguished as genera. Consequently, those names of taxa formally associated with the genus category should no longer be italicized (de Queiroz and Gauthier 1992, 473). They should be treated in the same manner as the names of other higher taxa, merely capitalized. Instead of "*Homo*" we should write "Homo." This change in nomenclatural practice is fairly simple. Reforming the current use of binomials is more complicated.

Binomials are used by the current system of nomenclature to indicate which taxa are assigned the rank of species. But if Linnaean ranks are dropped, then so too should the nomenclatural distinction between species and higher taxa. How, then, should a post-Linnaean system of nomenclature treat already existing binomials? Three proposals are found in the literature: (1) preserve current binomials; (2) hyphenate the generic and specific names of a binomial to form a uninomial; (3) run generic and specific names together and form a uninomial. Let us consider each proposal.

Griffiths (1976, 172) advocates the preservation of current binomials. He argues that if we eliminate the generic names found in current binomials, then the resulting uninomials will give rise to many homonyms. Such homonymy can be avoided, but only at the cost of renaming many taxa. Better to allow classifications to have different types of names (binomials and uninomials) than widespread homonymy or the wholesale renaming of taxa. According to Griffiths's proposal, a taxon's binomial would not indicate its level of inclusiveness. The difference in types of names would be just a historical remnant of earlier nomenclatural rules. One implication of Griffiths's suggestion is that binomials would be found at various levels of inclusiveness within a classification, not just at basal levels. Suppose a taxon initially thought to be a basal taxon and assigned a binomial is found to be a more inclusive taxon; its binomial would remain. In a post-Linnaean system, the classification of the taxon changes but not its name.

Binomials in the Linnaean system not only indicate the ranks of taxa but also their taxonomic placement: the generic name in a binomial indicates a species' genus; and if a species is assigned to a new genus, its generic name must be altered. Griffiths wants to promote stability by avoiding the renaming of taxa. He calls the first name of a binomial a *forename* rather than a generic name, and he proposes that forenames, unlike generic names, remain constant even when a taxon with a binomial is assigned to a different next inclusive taxon. For example, the taxon *Paracolletes friesei* was found to be a part of the taxon *Leioproctus* rather than *Paracolletes*. On Griffiths's suggestion, that taxon should retain its original name, *Paracolletes friesei*, rather than be renamed *Leioproctus friesei*.

While Griffiths promotes the continued use of binomials, Cain (1959a, 242–3) and Michener (1963, 164–5) push for replacing binomials with uninomials. They offer two options. In one, Cain and Michener suggest shortening the generic name of a binomial to a single letter, then hyphenating that letter to the binomial's specific name. The result is a hyphenated uninomial. For example, *Homo sapiens* would become *H-sapiens*. In the other proposal, Michener (1963, 164–5) suggests taking current binomials and running their generic and specific names together. *Homo sapiens* would become *Homosapiens*. Cain and Michener's motivation for incorporating the generic name of a taxon (or parts of a generic name) within a new uninomial is much the same as Griffiths's reason for preserving binomials: the switch from binomials to uninomials of just specific names would cause too many homonyms. The preservation of generic names, or parts thereof, helps avoid homonymy. The avoidance of homonymy, in turn, avoids the introduction of new names that would cause instability.

Which one of these three proposals for dealing with current binomials in a post-Linnaean system is preferable? There are pros and cons for each. The third option, running together the parts of binomials to form uninomials, might be the most ideal. If that proposal were adopted, no distinctions among the names of taxa would exist: all taxa would have uninomials and none of those uninomials would contain hyphens. The assumption that a difference in taxon name indicates an ontological difference would be avoided. But there are practical problems with this proposal. For one, the names of such uninomials might be uncomfortably long. Consider the name *Euschongastiannmerica* (Michener 1963, 165). Furthermore, the change from binomials to uninomials would require the altering of many names. All binomials in all classifications would need to be altered. The theoretical desire to have a uniform name type for all taxa does not outweigh the inconvenience of adopting such a system.

The next ideal proposal would be to form uninomials by taking the initials of generic names and hyphenating them with specific names. Again, there would be no misleading mix of uninomials and binomials in a post-Linnaean classification. But like the previous proposal, this one has problems. Would using just the first letter of a generic name in forming a uninomial be sufficient for avoiding homonymy? More important, the newly formed uninomials would be strikingly different from their former binomials; they would be just a single letter hyphenated to a specific name. Biologists would then need to learn the new names of all basal taxa they study. This is no small task, and it would cause the sort of instability that taxonomic systems, since Linnaeus's, are designed to avoid. Finally, a more theoretical problem faces this proposal. The change from binomials to uninomials was motivated by the belief that all taxon names should have the same form. Uninomials with hyphens are different from uninomials without hyphens. Therefore, this proposal fails to meet the overall objective for switching to uninomials.

Perhaps, then, the most practical option is to preserve currently used binomials. That option has the very practical virtue of not requiring biologists to relearn the names of all base taxa. As a result, this proposal provides the most stability in a switch from a Linnaean to a post-Linnaean classification. Of course, this proposal, as any, has its problems. If binomials are preserved, then post-Linnaean classifications will consist of two types of names, uninomials and binomials. In such classifications, having a binomial versus a uninominal would not indicate the level of inclusiveness of a taxon. Binomials might refer to basal taxa or they might refer to more inclusive taxa. That might be confusing because the occurrence of a binomial might mislead one to think that a taxon is less inclusive than it is.

However, the potential for such confusion is not as strong as it may appear at first. Biologists and others can certainly learn that the occurrence of a binomial in a post-Linnaean classification does not indicate the taxonomic position of a taxon. Consider a parallel situation in everyday life. Some people have two names, some have three, some have more; yet these differences do not cause us to think that people with different numbers of names vary accordingly. Similarly, the intermixing of uninomials and binomials in post-Linnaean classifications should not cause us to think that taxa with different name forms are different types of taxa. The appearance of binomials in post-Linnaean classifications is merely a historical artifact of the Linnaean system.

One last point should be addressed concerning the continued use of binomials. Should current binomials remain italicized in a post-Linnaean system? As we saw, de Queiroz and Gauthier address this question concerning those taxa previously assigned the rank of genus. They suggest dropping such italicization so as to avoid the existence of inappropriate nomenclatural devices. By preserving the use of binomials in a post-Linnaean system we are keeping an inappropriate nomenclatural device. But this is forced upon us for pragmatic and historical reasons. Nevertheless, we can clean up our act a little bit and drop the italicization of binomials.

So far we have addressed the question of what to do with those taxa that already have binomials. We have not yet considered whether newly discovered basal taxa should be given binomials or uninomials. Griffiths (1976, 172) argues that such taxa should be given binomials, whereas Michener (1963, 165) suggests that they should be given hyphenated uninomials. Neither promotes the use of pure uninomials for newly found basal taxa. Their motivation is similar to the one we saw earlier: so many basal taxa exist that the assignment of forenames to newly discovered basal taxa would help avoid the occurrence of homonymy.[2]

Suppose we adopt the proposal that newly discovered basal taxa should be assigned binomials. How should the parts of these binomials be assigned? Specific names, in traditional nomenclature, do not indicate the taxonomic position of a taxon but are chosen to avoid homonymy. Generic names are chosen to indicate the taxonomic placement of a species. In a post-Linnaean system, would the forename of a taxon be the name of its next inclusive taxon? Forenames could be chosen in that manner, but only when keeping in mind that forenames are merely names and not references to the taxonomic positions of taxa. Alternatively, forenames could be assigned in the same way that specific names are chosen, with no reference to any other taxa. The important point is not to embody forenames with taxonomic significance – they are merely parts of names.

Instead of assigning binomials to newly discovered basal taxa, they could be given pure uninomials (that is, uninomials lacking hyphens). Neither Michener nor Griffiths opt for this suggestion; they argue that because so many basal taxa exist, forenames are needed to avoid homonymy. But the occurrence of homonymy does not force us to assign binomials to newly discovered basal taxa. Surely, suitable uninomials that refer to just one taxon can be devised: there is virtually no limit on which uninomials can be formed. So the threat of homonymy does not imply that we should create more binomials.

A post-Linnaean system of nomenclature needs to balance two competing concerns: the avoidance of nomenclatural devices that inappropriately signal a difference between taxa, and minimizing, as much as possible, the disruption caused by adopting new rules of nomenclature. Newly discovered basal taxa do not yet have names, so assigning them uninomials would not cause the sort of instability that occurs when current binomials are replaced with uninomials. Newly found basal taxa should be assigned uninomials. To do otherwise is to adopt nomenclatural differences at the start when there is no practical reason for doing so. Having to live with preexisting binomials in a post-Linnaean system is a condition that arises from pragmatic concerns and the historical fact that binomials have been assigned for many years. Newly discovered basal taxa need not be saddled with the remnants of a previous system.

One last traditional nomenclatural device remains to be discussed, the use of trinomials for subspecies and varieties. In the Zoological and Botanical Codes, such taxa are named with three terms. In the Bacteriological Code, such taxa have a word indicating their rank inserted in their names, for example, the "subsp." in *Bacillus subtilis* subsp. *Niger* indicates that it is a subspecies (Jeffery 1989, 12). The treatment of trinomials in a post-Linnaean system follows the path already discussed. Current trinomials should be preserved to promote stability. Words explicitly referring to such Linnaean ranks as subspecies or variety should be dropped. In other words, the resultant names should no longer indicate the rank or taxonomic position of a taxon.

What about newly discovered subspecies and varieties? Whether such taxa should be formally recognized is a controversial issue. Rules of nomenclature should be neutral on this issue, and in being so they should not foreclose the possibility that such units could be recognized. If subspecies and varieties are recognized, then they should be assigned uninomials. Again we follow the idea that the taxonomic level of a taxon, or its placement, should not be written in its name.

Let us step back and summarize the recommendations offered in this section.

Recommendations

R3. Suffixes in already existing taxon names are kept, but they no longer indicate categorical rank.
R4. The inclusion of root names in newly formed names is optional. If a root is used, it does not indicate a taxon's placement.
R5. The assignment of suffixes to newly formed names is optional. If a suffix is used, it does not indicate a taxon's rank.
R6. Taxon names are no longer italicized.
R7. Binomials already in use are kept to maintain stability; but a binomial no longer indicates the rank or placement of a taxon.
R8. Newly discovered basal taxa are assigned uninomials.
R9. Trinomials already in use are kept to maintain stability; but a trinomial no longer indicates the rank or placement of a taxon.
R10. Newly discovered taxa thought to be less inclusive than basal taxa are assigned uninomials.

7.3 DEFINING TAXON NAMES

Another aspect of biological nomenclature needs to be addressed: How do taxon names refer to taxa? The semantic glue between a taxon name and its referent is a name's definition. Linnaeus assumed that taxon names are defined by the essential natures of taxa. The contemporary rules of nomenclature have rejected essentialism and now define taxon names according to the type method. Although the type method distances itself from essentialism, de Queiroz and Gauthier argue that the contemporary codes do not go far enough (de Queiroz and Gauthier 1990, 1992, 1994; de Queiroz 1992, 1994). In its place they suggest that taxon names be given phylogenetic definitions. Much of this section will explore de Queiroz and Gauthier's alternative system for defining taxon names. But first some background information is in order. We will take a quick look at Linnaeus's method for defining taxon names as well as the method offered in the contemporary codes.

Linnaeus is very clear on how taxon names should be defined: by listing those characteristics that are common and unique to the members of a taxon. The definition of a specific name, for example, highlights "some characteristic which is not found in any other plant of the genus" (Linnaeus 1731[1938], section 290). Generic names are defined by highlighting the distinctive marks of a genus's fructification system (Larson 1971, 122–3). The names of classes, orders, and varieties are defined in terms of the essential characters of those taxa (Larson 1971, 139). Linnaeus believed that biologists should strive to provide essentialist definitions of taxon names (1731[1938], sections 288–90), but he was fully aware that such definitions often cannot be given. In their place, and hopefully as a temporary measure, Linnaeus suggested that we provide "synoptic" definitions. Synoptic definitions list those characteristics common to several species along with a dichotomous classification of those taxa.

The notion that the organisms of a taxon share common and unique characteristics no longer has a home in biology (Chapter 3). From an evolutionary perspective, species are historical entities consisting of appropriately connected organisms. They are not classes of entities sharing qualitative essences. The contemporary codes of nomenclature have been written in light of evolutionary biology. They make no essentialist assumptions but instead use the type method for defining taxon names. Accordingly, the definition of a taxon name refers to the nomenclatural type of a taxon, not the traits found among that taxon's organisms.

What, then, are nomenclatural types? They vary among codes. According to the Zoological Code, the type for a family or higher-ranked taxon is a genus, the type for a genus is a species, and the type for a species is a specimen. How are types used to define taxon names? "The type is the objective basis to which a given name is permanently linked and it is by the type method that the correct application of names is objectively and unequivocally determined" (Jeffrey 1989, 21). An example would help. Suppose we want to define the name of the family Tipulidae. The name of that taxon is defined as "the taxon containing the type genus *Tipula* that is assigned the rank of family" (de Queiroz and Gauthier 1994, 28). By definition, the name "Tipulidae" sticks to the family with that genus. Suppose Tipulidae is found to contain many genera and is divided into two families. Given the definition of "Tipulidae," the family with that name must contain the genus *Tipula*. A taxon name must refer to the taxon containing the type mentioned in that name's definition.

Types are often called "nominal types" to signify that they need not be typical in any qualitative sense. The characteristics found among the members of a species can vary at a time and over time, so much so that the characteristics of a type specimen may vary dramatically from those found in the rest of the species (Hull 1976[1992], 307). Types are not even given with standards for allowable variation within taxa (Wiley 1981, 391; Jeffrey 1989, 22). In fact, no qualitative traits are cited in a type-based definition. Such definitions only cite a type taxon or a type specimen and the rank of the taxon whose name is being defined.[3]

If type-based definitions merely mention the name of a type taxon and none of the characteristics found among the organisms of that type, how does a type serve as a means for connecting a taxon's name to a taxon? One suggestion is that taxon names are ostensively defined (Ghiselin 1974; Hull 1978[1992]). Taxon names are proper names, and proper names are defined by pointing to the entity in question, or a part of it, and uttering a name. Someone might ostensively define the name "Ereshefsky" by pointing to me and uttering "the name 'Ereshefsky' refers to that person over there." In the case of taxa, a biologist rarely has an entire taxon in front of her, so when ostensively defining a taxon name she points to a type specimén or utters the name of a type taxon. A species name, for example, is attached to a species by pointing to a specimen of that species and uttering that species' name. When ostensively defining the name of a higher taxon a biologist does not point to a speciman but utters the name of an appropriate type taxon. For example, the name "Tipulidae" is ostensively defined by uttering "the name 'Tipulidae' refers to the family taxon containing the type *Tipula*."

Ostensive definitions are an important improvement over essentialist ones. Ostensive definitions escape the pitfalls of essentialism by defining taxon names in terms of type specimens and type taxa rather than the qualitative features of organisms. Nevertheless, ostensive definitions have their problems. For one, such definitions explicitly refer to the Linnaean categories and in doing so bring in questionable assumptions concerning the nature of those categories. In the example above, "Tipulidae" is defined as the *family* taxon containing *Tipula*. The reference to the category "family" assumes that we have a prior conception of family; in particular, it assumes that we have a conception of how families differ from other types of taxa. Without a prior conception of family the ostensive definition of Tipulidae is ambiguous: it can refer to any number of taxa containing *Tipula*. An important claim of Chapter 6 is that the Linnaean ranks no longer have any general meaning beyond indicating the relative level of inclusiveness of a taxon within a specific group of taxa. So ostensive definitions using types rely on the problematic assumption that the Linnaean ranks have theoretical significance.[4] We have seen that the type method for defining taxon names nicely drops the assumption that individual taxa have essences. But ostensive definitions do not go far enough because they still retain the assumption that the Linnaean ranks are ontologically meaningful designations.

De Queiroz and Gauthier (1990, 1992, 1994) point to another problem with ostensive definitions. An ostensive definition mentions that the higher taxon being named contains a certain type genus, but it says nothing further about the biological relation between the taxon named and the type. According to de Queiroz and Gauthier (1992, 460), "it would be useful to state the definition of each taxon name in terms of a specified relationship to a type." This suggestion makes eminent sense. De Queiroz and Gauthier merely ask that we make explicit what is already assumed in a type definition, the hypothesized evolutionary relations between the taxon being named and its type. Definitions that explicitly state such relations are not incompatible with ostensive definitions but supplement them (de Queiroz and Gauthier 1992, 460; also de Queiroz 1992).

Still, de Queiroz sees a striking problem with ostensive definitions: they fail to provide necessary and sufficient conditions for applying names. In his survey of methods for defining taxon names, de Queiroz writes the following.

> In order to accommodate the possibility of evolutionary change, either the lists of defining characters had to be made less definite, or else definition in terms of necessary and sufficient properties had to be abandoned altogether. Neither of those alternatives was able to provide satisfactory definitions of taxon names (1992, 299–300).

The first alternative – providing lists of defining characters that are less definite than the traditional approach – is the cluster method. We discussed the cluster method and its problems in Section 3.2. The second alternative – abandoning the use of necessary and sufficient properties – is the method of ostensive definition. According to de Queiroz, a problem for both methods is their inability to provide precise criteria for when to use a taxon name. In de Queiroz's terminology, they do not provide definitions "in terms of necessary and sufficient properties." In fact, for de Queiroz, ostensive definitions fail to be definitions at all:

> Hull (1976) considered ostensive definitions analogous to baptismal acts, which are not definitions in the traditional sense; consequently, he proposed that the proper names of individuals cannot be defined at all (1992, 302).

Hull (1976), as far as I know, does not assert that ostensive definitions are not definitions. This is de Queiroz's assertion. Again, the problem is that ostensive definitions fail to provide necessary and sufficient conditions for the application of a taxon name. We will return to this criticism of ostensive definitions shortly, but first let us see de Queiroz and Gauthier's positive proposal for defining taxon names.

De Queiroz and Gauthier (1990, 1992, 1994) suggest replacing ostensive definitions with "phylogenetic definitions."[5] Definitions of the names of monophyletic taxa can be given in three ways. A *node-based definition* is used to define the name of a clade stemming from the most recent common ancestor of two other taxa (Figure 7.2a). For example, "Leopidosauria" is defined as "the most recent common ancestor of *Sphenodon* and squamates and all of their descendants" (de Queiroz

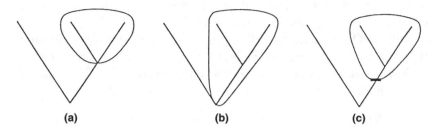

Figure 7.2   Phylogenetic definitions of taxon names. (a) A node-based definition defines the name of a taxon as the taxon stemming from a most common
recent ancestor. (b) A stem-based definition defines the name of a taxon as all
those species sharing a more recent common ancestor with one taxon than with
another. (c) An apomorphy-based definition defines the name of a taxon as the
taxon stemming from the first ancestor to have a particular synapomorphy.
Adapted from de Queiroz and Gauthier 1990, 310.

and Gauthier 1990, 310). A *stem-based definition* defines the name of a
clade consisting of all those species that share a more recent common
ancestor with one taxon than another (Figure 7.2b). Accordingly,
"Lepidosauromorpha" is defined as "Lepidosauri and all species sharing a more recent ancestor of Lepidosauri than with Archosauria" (de
Queiroz and Gauthier 1994, 29). An *apomorphy-based definition* is used
to define the name of a taxon stemming from the first ancestor to have
a particular synapomorphy (Figure 7.2c). For example, "Tetrapoda" is
commonly defined as "the first vertebrate to possess digits and all its
descendants" (de Queiroz and Gauthier 1990, 310).

   Phylogenetic definitions are introduced with the intention of defining the names of monophyletic taxa. Nevertheless, de Queiroz and
Gauthier (1990, 311) allow that phylogenetic definitions can also be
used to define the names of paraphyletic and polyphyletic taxa. For
instance, "Reptilia" could be given the stem-based definition "the most
recent common ancestor of Mammalia and Aves and all its descendants
except Mammalia and Aves." The possibility of providing phylogenetic
definitions for nonmonophyletic taxa is a positive attribute of their
system. Still, de Queiroz and Gauthier maintain the cladistic position
that eventually all references to paraphyly and polyphyly should be
eliminated from taxonomy, save, perhaps, when one is talking about the
history of taxonomy.

   De Queiroz and Gauthier argue that phylogenetic definitions have
a number of advantages over type-based definitions. A major flaw that
they see with the current system of nomenclature is that "the defini-

tions of taxon names are stated in terms of characters, that is, organismal traits. . . . Definitions of taxon names are fundamentally nonevolutionary" (de Queiroz and Gauthier 1992, 460). Phylogenetic definitions, on the other hand, "are stated in terms of common descent and phylogenetic entities" (de Queiroz and Gauthier 1994, 28). When de Queiroz and Gauthier assert that definitions citing traits are nonevolutionary, they mean one of two things. Either such definitions permanently assign certain traits to the members of a taxon such that the taxon cannot evolve; in other words, all organisms in that taxon, past, present, and future, must have that set of traits. Or de Queiroz and Gauthier mean that any definition citing a qualitative trait is nonevolutionary. Here they would be following the cladistic line that only relations should be cited in defining taxon names.

On either reading, ostensive definitions are not vulnerable to the charge of being nonevolutionary. As we have seen, the type specimens, or type taxa, cited by ostensive definitions need not be typical (Jeffery 1989, 21–2; Wiley 1981, 391). The members of a taxon can be quite different from the type specimen used to define that taxon's name. Moreover, ostensive definitions do not list any organismic traits, let alone those that the members of a taxon must have; they merely mention a type specimen or type taxon. So phylogenetic definitions are not the only sorts of definitions that avoid citing organismic traits.

Nevertheless, one attribute that sets phylogenetic definitions apart is their explicit reference to evolutionary relations. Recall what an ostensive definition under the type method looks like. The definition of "Tipulidae" is "the taxon containing the type genus *Tipula* that is assigned the rank of family." What binds the type genus *Tipula* to the other parts of Tipulidae are various evolutionary relations. Such relations are not stated but only implicitly assumed in ostensive definitions. Phylogenetic definitions make those relations explicit – phylogenetic definitions wear the evolutionary nature of taxa on their sleeves. Such explicitness is nice in itself, but it is particularly helpful in avoiding confusion. Because ostensive definitions do not state which relations bind the parts of a taxon, some confusion may arise over the nature of a taxon. For example, the ostensive definition of "Tipulidae" tells us that it is the family-ranked taxon containing the genus *Tipula* but it does not tell us whether Tipulidae is monophyletic or paraphyletic. Phylogenetic definitions make such information evident.

De Queiroz sees another, more philosophical attribute of phylogenetic definitions. Linnaean definitions were designed to provide necessary and sufficient conditions in terms of organismic traits. But as we know, no organismic traits are essential for membership in a taxon. Ostensive definitions avoid the pitfalls of essentialism by making no reference to organismic traits. Still, de Queiroz (1992, 300) finds ostensive definitions unsatisfactory because they fail to provide definitions in terms of necessary and sufficient conditions. That is, ostensive definitions fail to give precise conditions for when, and only when, a taxon's name should be applied. Phylogenetic definitions are supposed to do better because they avoid the pitfalls of Linnaean definitions yet provide precise conditions for when a taxon name should be used. The necessary and sufficient conditions that phylogenetic definitions cite are particular facts about ancestry (de Queiroz 1992, 300, 304). For example, using a node-based definition of "Leopidosauria," we can say that an organism is a member of Leopidosauria if and only if it is a descendant of the most recent common ancestor of *Sphenodon* and squamates or it is a member of that ancestral taxon.

Phylogenetic definitions are certainly a step in the right direction. They explicitly cite those evolutionary relations that bind organisms and subtaxa into single biological units, and they draw our attention away from defining taxon names in terms of qualitative characters. Furthermore, they drop any reference to Linnaean categories in defining the names of taxa. However, phylogenetic definitions do not fulfill the philosophical desideratum that de Queiroz places on scientific definitions: they do not provide complete definitions of taxon names. Phylogenetic definitions are not alone in this regard; no empirically sensitive definition of a taxon name can live up to such philosophical aspirations. Complete definitions of taxon names are not forthcoming because taxa are not entities with precise ontological boundaries. The rest of this section illustrates this point by using phylogenetic definitions, but the lesson here is a general one for biological nomenclature: given the ontological nature of taxa, we should not aspire to semantic essentialism in our definitions of taxon names.[6]

Ancestors play an important role in phylogenetic definitions. Both node-based definitions and stem-based definitions refer to ancestors. However, the existence and classification of ancestors is a controversial matter for cladists (Hull 1979, Ridley 1986). That controversy rears its head in phylogenetic definitions. Consider the node-based definition of "Leopidosauria": "the most recent common ancestor of *Sphenodon* and squamates and all of their descendants." "Leopidosauria" is defined by reference to three other taxa: *Sphenodon*, squamates, and their

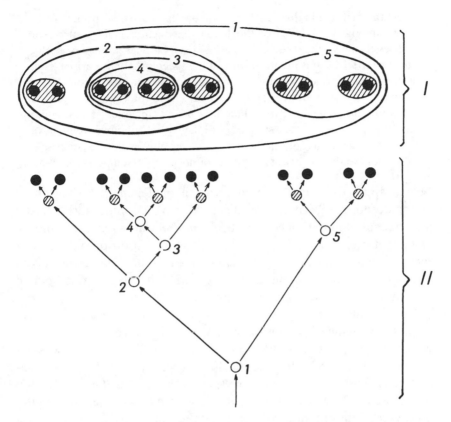

Figure 7.3   Diagram I illustrates the part-whole relations of taxa 1 through 5. Diagram II shows their phylogenetic relations. From W. Hennig, *Phylogenetic Systematics*. Copyright 1966, 1979 by the Board of Trustees of the University of Illinois. Used with the permission of the University of Illinois Press.

most recent common ancestor. For this definition to provide necessary and sufficient conditions for the application of "Leopidosauria," the taxon names offered in the definition must be defined as well. But how do we define "the most recent common ancestor of *Sphenodon* and squamates"? More to the point, what is that ancestor? Here I am not asking an empirical question but a methodological question: How are ancestors conceived in cladism?

One suggestion from Hennig is that an ancestor is equivalent to the group containing all of that ancestor's descendant taxa. Consider his discussion of stem species and his accompanying diagram (Figure 7.3):

From the fact that in diagram 1 the boundaries of a "stem species" coincide with the boundaries of the taxon that includes all its successor species, it follows that the "stem species" itself belongs in this taxon. But since, so to speak, it is identical with all the species that have arisen from it, the "stem species" occupies a special position in this taxon. If, for example, we knew with certainty the stem species of the birds . . . , then we would no doubt have to include it in the group "Aves." But it could not be placed in any of the subgroups of the Aves. Rather we would have to express unmistakably the fact that in the phylogenetic system it is equivalent to the totality of all species of the group (Hennig 1966, 71–2).

Let us apply this suggestion in giving a phylogenetic definition of "Leopidosauria." The node-based definition of "Leopidosauria" is "the most recent common ancestor of *Sphenodon* and squamates and all of their descendants." Yet, according to Hennig's proposal, the most recent common ancestor of *Sphenodon* and squamates is the taxon Leopidosauria. So equating an ancestor with the taxon being defined renders phylogenetic definitions circular.

Such circularity can be avoided if an ancestor is merely a part of the taxon being defined and not the entire taxon itself. This solution might work, but then there is the problem of recognizing such ancestors without violating the principle that all taxa are monophyletic. One suggestion is to use only interbreeding species as ancestors (Ax 1987, 31–3; de Queiroz 1988, 255–6). The existence of such species depends on the process of interbreeding rather than the relationships of common ancestry. Although interbreeding species fail to be monophyletic, they are nevertheless considered real entities because of the interbreeding relations that bind them into single units (de Queiroz 1988, 256; de Queiroz 1992, 304).

Whether interbreeding species can be real nonmonophyletic ancestors is yet another controversial matter (de Queiroz and Donoghue 1988). I want to put that controversy to one side and instead ask: If such species exist, could they then be used to provide complete phylogenetic definitions? Two problems stand in the way. First, many higher taxa do not contain interbreeding populations as their basal units. Higher taxa consisting of asexual organisms immediately come to mind. If we think that higher taxa consisting of asexual organisms deserve names, then a method that defines taxon names only in terms interbreeding species will leave out much of life on this planet.

A more philosophical problem stands in the way of providing complete definitions. If we want to give complete definitions of taxon names, then we need the terms in those definitions to be complete. That is, the terms within such definitions must themselves be defined by necessary and sufficient conditions. But rarely, if ever, can we do that for any definition of a population. (Here I am assuming that populations owe their existence to the relations among their members rather than their sharing qualitative similarities.) Consider interbreeding populations. Such populations have vague boundaries, both synchronically and diachronically (Splitter 1988). Hybrid zones and ring species contain organisms that are not clearly in one species or another. Similarly the boundaries between species during speciation are often vague, so much so that no determinate answer can be given whether an organism is a member of one interbreeding population or another. As we saw in Section 1.3, such borderline cases are not ones where if we were provided with more information we could assign an organism to a species. The issue here is not epistemological, but the ontological nature of population boundaries – they are naturally vague.

The bearing on definitions of taxon names is this. We cannot give complete and precise definitions of taxon names because the terms used in those definitions, namely, those that refer to populations, are themselves incomplete and imprecise. This problem does not stem from a particular school of biological taxonomy because it arises in any method that relies on defining taxon names with reference to populations. The problem is with the philosophical assumption that the definitions of taxon names should be complete, that we should aspire for definitions of taxon names that provide necessary and sufficient conditions. That assumption is simply too much to ask of the definitions of taxon names.

Alternatively we could require that a definition works in the vast majority of cases yet allow that it fails in borderline cases. So, for example, a node-based phylogenetic definition will cite an ancestor that can be used to locate the vast majority of subtaxa within the taxon defined. Yet this definition will not be considered a failed one when borderline cases are encountered. Perhaps some subtaxa and organisms are not clearly parts of the taxon being defined as the result of introgression. Perhaps the boundaries of the ancestral population of the taxon are vague. Such borderline cases do not undermine the theoretical and practical merit of a definition unless they become widespread. If that occurs, then a definition may need to be recast or rejected.

Let us step back and take stock. In our review of the future of Linnaean nomenclature we came to the issue of defining taxon names. We saw that the traditional Linnaean approach requires that the members of a taxon share a set of common essential characteristics, but that requirement is inconsistent with evolutionary theory and should be rejected. Ostensive definitions, on the other hand, cite type taxa and type specimens rather than the qualitative traits of organisms, so they avoid the problems facing essentialist definitions. Recently de Queiroz and Gauthier have argued that we could do better than ostensive definitions – we should provide definitions of taxon names that explicitly state the evolutionary relations that bind the parts of a taxon. Phylogenetic definitions do just that, and in doing so, they are a step in the right direction. Furthermore, phylogenetic definitions have the virtue of defining taxon names without any reference to the Linnaean ranks. However, the introduction of phylogenetic definitions is overlaid with the philosophical aspiration that adequate definitions be complete, that definitions provide necessary and sufficient conditions for the application of taxon names. But no definitions of taxon names, phylogenetic or otherwise, can satisfy those requirements. Taxa and populations have natural vagaries, so they are not the sorts of entities that are amenable to precise definitions. We have rejected Linnaeus's essentialist definitions based on characters. We should not allow essentialism to rear its head again in phylogenetic definitions.

Nevertheless, this conclusion should not stop us from using phylogenetic definitions. Phylogenetic definitions are a vast improvement over Linnaean definitions: they put some theoretical flesh on ostensive definitions, and they avoid the use of the Linnaean categories. Here, then, is another worthy change in nomenclatural practice, one that places further distance between the Linnaean system and the post-Linnaean system of classification.

Recommendation

R11. Where possible, taxon names should be given phylogenetic definitions. Such definitions can be used to define the names of monophyletic and nonmonophyletic taxa.

8

The Future of Biological Nomenclature

The last chapter introduced and analyzed a number of proposals for an alternative system of classification. The best parts of those proposals were then woven into a post-Linnaean system containing eleven recommendations. Those recommendations address three aspects of biological taxonomy: how to represent the hierarchical relations of taxa, how to name taxa, and how to define taxon names. With those recommendations in hand we can now compare the post-Linnaean system with the Linnaean system. But to do that we must decide which version of the Linnaean system should be used as a basis for comparison, for as Chapter 6 illustrates the Linnaean system has gone through more than one incarnation. To put the Linnaean system on its best footing, we should pick the best developed version of that system. Arguably, the most complete and up-to-date version is Wiley's (1979, 1981) "Annotated Linnaean Hierarchy."

This chapter contains a comparison of the annotated Linnaean system with the post-Linnaean system developed in the previous chapter. Section 8.1 introduces the annotated Linnaean system and many of its problems. Section 8.2 turns to the post-Linnaean system and examines whether the post-Linnaean system does a better job at overcoming those problems. Section 8.3 steps back from a detailed comparison of the two systems and considers a more general issue. Suppose the post-Linnaean system better coheres with evolutionary theory and, if adopted, would make the job of taxonomists easier. Suppose, in other words, that the post-Linnaean system is preferable for both theoretical and pragmatic reasons. At first glance such reasons seem sufficient for adopting a new system of nomenclature, but in practice they are not. Given the pervasiveness of the Linnaean system both in and outside of biology, one might wonder if the switch to an alternative system is practically feasible. Section 8.3 weighs the benefits of adopting the post-Linnaean system against the inconvenience of abandoning the current system.

8.1 THE ANNOTATED LINNAEAN HIERARCHY

Bits and pieces of Wiley's (1979, 1981) annotated Linnaean system were introduced and discussed in Section 6.3. However, a proper comparison of the annotated system with the post-Linnaean system requires the introduction of Wiley's entire system, so we will start with that. The annotated system contains two rules and nine conventions. Here they are, verbatim, from Wiley's 1981 text.

Rule 1. Taxa classified without qualification are monophyletic groups *sensu* Hennig (1966). Nonmonophyletic groups may be added if they are clearly qualified as such.

Rule 2. The relationships of taxa within the classification must be expressed exactly.

Convention 1. The Linnaean hierarchy will be used, with certain other conventions to classify organisms.

Convention 2. Minimum taxonomic decisions will always be made to construct a classification or modify an existing classification. This will be accomplished in two ways. First, no empty or redundant categories will be produced unless these categories are necessary taxonomic conventions (i.e., the five mandatory categories may be redundant). Second, natural taxa of essential importance to the group classified will be retained at their traditional ranks whenever possible, consistent with phylogenetic relationships and the taxonomy of the group as a whole.

Convention 3. Taxa forming an asymmetrical part of a phylogenetic tree may be placed at the same categorical rank (or designated "plesion" and indented) and sequenced in the order of origin. When an unannotated list of taxa is encountered in a phylogenetic classification, it is understood that the list forms a completely dichotomous sequence, with the first taxon listed being the sister group of all subsequent taxa and so on down the list.

Convention 4. Monophyletic groups in trichotomous or polytomous interrelationships will be given equivalent ranks and placed *sedis mutabilis* at the level of the hierarchy at which their relationships to other taxa are known.

268

Convention 5. Fossil or Recent monophyletic taxa of uncertain relationships will be placed in the hierarchy *incertae sedis* at the level their relationships are best understood.

Convention 6. A paraphyletic or polyphyletic assemblage, or group of unknown status may be included in a phylogenetic classification if it is put in shutter quotes that indicate that all of the included taxa are *incertae sedis* at the level in the hierarchy that the group is placed. Such groups will not be given rank (neither formal rank nor plesion status).

Convention 7. Fossil groups will be classified in a different way than Recent groups. The status "plesion" will be accorded to all monophyletic fossil taxa. When a plesion is sequenced in a combined Recent-fossil classification, it is the sister group of all the other terminal taxa within its clade and below it in that classification. Plesions may also stand *sedis mutabilis* or *incertae sedis* relative to other plesions and to Recent taxa.

Convention 8. A stem species of a suprageneric taxon will be classified in a monotypic genus and placed in the hierarchy in parentheses beside the taxon which contains its descendants. A stem species of a genus will be classified in that genus and placed in parentheses beside the generic name.

Convention 9. A taxon of hybrid origin will be classified with one or both parental taxa and its hybrid nature will be indicated by placing the names of its parental species in parentheses beside the hybrid's name. *The sequence of the hybrid taxon carries no connotation of branching relative to subsequently sequenced taxa of nonhybrid origins.*

These rules and conventions address a number of concerns in biological taxonomy. Many were discussed in Chapter 6; nevertheless, here is a quick review. Rule 1 and Convention 6 bring to the fore the cladistic belief that nonmonophyletic groups are not "natural." Accordingly, monophyletic taxa are classified without special designation, while nonmonophyletic taxa are labeled as such. Conventions 2, 3, and 7 are designed to limit the number of categories and taxa cited in classifications. The motivation for doing so is twofold: to prevent the expansion of the Linnaean hierarchy beyond its current twenty-one ranks, and to make taxonomic revision less onerous. In particular, Convention 2 tells us not to posit redundant categories other than the standard mandatory five categories – genus, family, order, class, and phylum. Convention 3 limits the number of categories and taxa cited in a classification by using phyletic sequencing (see Section 6.3). Convention 7 limits the number of Linnaean categories by assigning fossil taxa the rank of plesion in classifications combining recent and fossil taxa (see Section 6.3).

The remaining conventions address a range of issues. Convention 4 supplements Convention 3 by offering a method for representing trichotomous and polytomous relations in classifications using phyletic sequencing. Although cladists disagree over whether such relations should be represented in classifications (see Hull 1979 for discussion), Wiley believes that they should and he develops a method for doing so. Convention 5 suggests how taxa with uncertain relations should be classified. Conventions 8 and 9 offer methods for incorporating information that traditional Linnaean classifications fail to provide. More specifically, Convention 8 provides a way of indicating which species are the ancestors of other taxa (see Section 6.3), while Convention 9 offers a method for displaying hybrid relations (see Section 6.4).

The annotated Linnaean hierarchy contains a number of improvements over the traditional Linnaean system. Although the annotated system is written for cladists, many of its conventions address general problems of traditional nomenclature – for example, how to indicate hybrid relations. The annotated system represents progress in biological taxonomy, but does it go far enough? For example, how well does it address the problems facing the traditional system? In what follows we will see that the annotated system does not go far enough. A number of fundamental problems with the traditional system are carried over to the annotated system. Moreover, the annotated system has its own problems.

To start, the annotated system continues to use the ontologically vacuous categories of the traditional Linnaean system. As we have seen, there is no reason to believe that the higher taxa of a Linnaean rank are comparable across classifications (Section 6.3). A family in one classification may be nothing like a family in another classification. At best, knowing that a taxon is a family within a particular classification tells us that it is older and more inclusive than a genus and a species in *that* classification. Other than that information, there is no general meaning associated with the higher Linnaean categories. The species category is a bit different; it has more theoretical significance than the categories of higher taxa. Nevertheless, the species category may also be a heterogeneous collection of noncomparable entities (Chapter 4). If the taxa of a particular Linnaean category are not comparable on some significant parameter, then we should doubt that category's existence. If the existence of the Linnaean categories is doubtful, then we should question their use in biological taxonomy. The continued use of the Linnaean categories misleads us to think that taxa of a particular rank are comparable across classifications, and worse yet causes us to argue over the rank of a taxon when such arguments lack an objective basis for resolution (Section 7.1). So the first problem with the annotated Linnaean system is that it preserves the misleading and theoretically empty ranks of the traditional system.

Next we turn to the problems that arise from the traditional rules of nomenclature. The Linnaean ranks of taxa are incorporated in taxon names through the use of binomials and suffixes. Species have binomials and the names of higher taxa have rank-specific endings. In addition, the taxonomic positions of species are indicated by their generic names, and the positions of some taxa above the level of genus are indicated by the roots of their names. As illustrated in Section 6.4, the incorporation of a taxon's rank and placement in its name leads to a number of problems. The annotated Linnaean system offers no changes to the traditional system's rules of nomenclature. Consequently, all of the nomenclatural problems of the traditional system are carried over to the annotated system.

Here is a quick review of the naming problems that affect both the traditional and annotated Linnaean systems. First, the ontological problems besetting the Linnaean ranks undermine the Linnaean rules of nomenclature. If the existence of the Linnaean categories is suspect, then so too is the practice of incorporating those categories in the names of taxa. If there are no species or genera, then no taxon should be designated as a "species" or a "genus." But let us suppose otherwise; let us suppose that the Linnaean categories are ontologically respectable. Still, a host of naming problems beset the traditional and annotated systems. The Linnaean requirement that a taxon's name indicates its position causes unnecessary instability and synonymy. Instability occurs when a taxon must be assigned a new name because it is found to have a different rank or position. Synonymy occurs because biologists are required to assign different names to a taxon when they disagree over its position. The source of such instability and synonymy is the Linnaean rules of nomenclature, not the ontological nature of taxa or our epistemological access to their positions in the world. (Section 6.4.) Consequently, Linnaean rules of nomenclature render the job of the taxonomist harder than necessary. Moreover, those Linnaean rules no longer achieve their desired effect; for example, binomials no longer serve as devices for memorizing the positions of all species within a kingdom (Section 6.2). Despite these problems, the annotated Linnaean system contains no amendments to the traditional rules of nomenclature but adopts them in their entirety.

The two types of problems just sketched – the maintenance of theoretically empty categories, and the use of pragmatically flawed rules of nomenclature – are significant problems for the annotated Linnaean system. They strike at the heart of the annotated system and seriously undermine its effectiveness. The annotated Linnaean system has other problems as well. These problems stem from that system's non-Linnaean recommendations (Ax 1987, Gauthier et al. 1988, de Queiroz and Gauthier 1992, Ereshefsky 1997).

Consider Conventions 3 and 7 of the annotated system. Convention 3 recommends the use of phyletic sequencing, and Convention 7 suggests that some taxa be given the non-Linnaean rank of plesion. Both conventions are offered as ways of stemming the proliferation of Linnaean ranks. Classifications using phyletic sequencing list the names of taxa according to their order of branching on asymmetrical cladograms. In doing so, such classifications avoid the positing of a number of inclusive taxa and their corresponding Linnaean ranks (see the example in Section 6.3). The use of plesion avoids the proliferation of Linnaean ranks in a simpler manner. Newly discovered fossil taxa are assigned the rank of plesion but are not assigned any Linnaean rank. As a result, the number of Linnaean ranks used in classifications combining recent and fossil taxa is cut down because many taxa have no Linnaean rank (see the example in Section 6.3).

Phyletic sequencing and plesions are introduced with the intention of overcoming certain flaws in the traditional system; however, their employment is not without problems. As illustrated in Chapter 6, the conventions of phyletic sequencing and plesion ranking are not amendments to the traditional system but significant departures from it. In an attempt to preserve the Linnaean system, these conventions require that we classify taxa in non-Linnaean ways. For example, when taxa are classified using phyletic sequencing, some inclusive taxa are neither classified nor given Linnaean ranks. Furthermore, those taxa phyletically sequenced have their hierarchical relations indicated by non-Linnaean means. When it comes to the use of plesions, fossil taxa assigned the rank of plesion are not given any Linnaean ranks. Therefore, in an attempt to avoid the proliferation of categories and the expansion of the Linnaean hierarchy, these conventions abandon two central tenets of the traditional system: that taxa be assigned Linnaean ranks, and that the hierarchical relations of taxa be indicated by those ranks. Wiley's conventions, in other words, preserve the Linnaean system by not applying it in numerous cases.

This is a philosophical observation, but it has a practical implication. As more taxa are classified using plesions and phyletic sequencing, biological classifications become less and less Linnaean. Such classifications may even become dominated by taxa classified with non-Linnaean means. A major aim of the annotated system is to preserve the use of the Linnaean system, but the annotated system's departure from the traditional system may counter that aim. The annotated system's conventions may even hasten the demise of the Linnaean system as biologists are forced to classify taxa in non-Linnaean ways.

The annotated Linnaean system fails to achieve its stated aims in other ways. One of Wiley's major reasons for promoting the annotated system is to reconcile phylogenetic systematics with the Linnaean system. Wiley believes that species taxa should be assigned the five mandatory categories of the Linnaean hierarchy (Convention 1). He also believes that classifications should have the ability to cite ancestral taxa (Convention 8). The rub is that he would like these conventions to be satisfied within the context of phylogenetic systematics. As Griffiths (1973, 1974), de Queiroz and Gauthier (1992, 455–7) and others note, this cannot be done. Here we return to the problem that ancestral taxa pose for cladists. In an extended quotation in Section 7.3 we saw how Hennig (1966, 71–2) characterized the problem: in the phylogenetic system, a stem species of a higher taxon is not merely a part of that higher taxon, it is identical to that higher taxon. Using Hennig's example, the stem species of Aves must be identical to Aves; otherwise, it would fail to be monophyletic.

Suppose we want to classify a stem species using Wiley's convention of five mandatory ranks. Then that species must be assigned a genus, a family, an order, a class, and a phylum. Given Hennig's position on stem species, that species is identical to a genus, a family, an order, a class, and a phylum. The problem for the annotated system is that a cladistic application of the five mandatory categories renders those categories meaningless: they apply to one and the same taxon and as a result refer to only a single hierarchical level. So a cladist has two choices. She can retain the Linnaean requirement that the five mandatory categories reflect differences in nature, but at the cost of allowing the existence of nonmonophyletic taxa. A stem species would be a part of a higher taxon rather than equivalent to that taxon. Or, she can maintain that only monophyletic taxa exist and reject the Linnaean assumption that the five mandatory categories reflect true divisions. Neither option allows the attainment of both Linnaean and cladistic ideals. Despite Wiley's hopes, the Linnaean system remains at odds with cladism.[1]

In sum, Wiley's annotated system faces a number of difficulties. First, its rules of nomenclature are those of the traditional Linnaean system. So all of the problems associated with traditional nomenclature are carried over to the annotated system. Second, the annotated system adopts the vacuous and misleading categories of the traditional system. And third, the annotated system fails to achieve two of its principal aims: the preservation of the Linnaean system, and the reconciliation of that system with cladism. In short, the annotated system fails to alleviate past problems of the Linnaean system while at the same time creating its own problems. The next question we consider is whether the post-Linnaean system developed in Chapter 7 can avoid these problems.

8.2   THE POST-LINNAEAN SYSTEM

Chapter 7 surveyed and analyzed a number of alternative systems of classification. The best parts of those systems were then put together to form a post-Linnaean system. Here are the eleven recommendations of the post-Linnaean system.

R1.   Taxa are no longer assigned Linnaean ranks.
R2.   The hierarchical relations of taxa within a classification are indicated by either indentations, positional numbers, or both.
R3.   Suffixes in already existing taxon names are kept, but they no longer indicate categorical rank.
R4.   The inclusion of root names in newly formed names is optional. If a root is used, it does not indicate a taxon's placement.
R5.   The assignment of suffixes to newly formed names is optional. If a suffix is used, it does not indicate a taxon's rank.
R6.   Taxon names are no longer italicized.
R7.   Binomials already in use are kept to maintain stability; but a binomial no longer indicates the rank or placement of a taxon.
R8.   Newly discovered basal taxa are assigned uninomials.

R9.   Trinomials already in use are kept to maintain stability; but a tri-
      nomial no longer indicates the rank or placement of a taxon.
R10. Newly discovered taxa thought to be less inclusive than basal
      taxa are assigned uninomials.
R11. Where possible, taxon names are given phylogenetic definitions.
      Such definitions can define the names of monophyletic and non-
      monophyletic taxa.

Recommendation 1 advocates eliminating the use of the Linnaean
ranks, and Recommendation 2 suggests alternative ways for representing hierarchical relations. Recommendations 3 through 10 provide
post-Linnaean rules of nomenclature. R11 promotes de Queiroz and
Gauthier's (1990, 1992, 1994) approach to defining taxon names. With
these recommendations in hand, let us see how the post-Linnaean
system handles the problems facing the annotated system.

To begin, the post-Linnaean system circumvents a number of problems associated with Linnaean ranks. As we have seen, taxa assigned
the same Linnaean rank often lack a common structure or common
degree of inclusiveness. Furthermore, the assignment of a taxon to one
rank rather than another often lacks empirical substantiation. Nevertheless, the continued use of those ranks misleads biologists and non-
biologists to think that taxa of the same rank are comparable, that the
assignment of a taxon to a Linnaean category is based on empirical
fact. The methods employed by the post-Linnaean system to indicate
the subordination of taxa carry no such misleading metaphysical connotations. Positional numbers and indentations merely indicate a
taxon's hierarchical position within a specific classification, nothing
more.

The use of the Linnaean ranks causes other problems for the traditional Linnaean system. Consider the problem of assigning multiple
ranks to a single taxon. Both annotated and post-Linnaean systems
address this problem, but the post-Linnaean system does so in a
simpler and more consistent fashion. Consider Wiley's Convention 2.
It recommends that the assignment of redundant ranks be avoided
unless those ranks are the five mandatory categories – species, genus,
family, order, and phylum. Using Wiley's (1981, 206) example, the subfamily Lepisosteinae is identical to the family Lepisosteidae, so the
name "Lepisosteinae" should be eliminated because its rank is redundant and not mandatory. In the same classification, the division Ging-
lymodi and the family Lepisosteidae are one and the same taxon, yet

according to Wiley we should not delete the name "Lepisosteidae" because it has a mandatory rank. The post-Linnaean system avoids the positing of redundant ranks in a simpler and easier manner – by not assigning ranks to any taxa. If no ranks are assigned to taxa, then no taxa are assigned multiple ranks. Furthermore, the post-Linnaean system addresses the problem of redundant ranks in a more thorough and consistent manner. The post-Linnaean system eliminates the existence of redundant categories at all levels, not just at some levels. The annotated system limits the introduction of redundant ranks in a piece-meal approach that treats some taxa (species, genera, families, orders, and phyla) one way and other taxa another way.[2]

Another area in which Wiley treats taxa unequally is the classification of fossil taxa. Convention 7 of the annotated system suggests that newly discovered fossil taxa be given the rank of plesion rather than a Linnaean rank. Recall, also, that the names of such taxa are inserted next to the names of their sister taxa (see Section 6.3 for a full discussion and an example). Wiley's motivation for Convention 7 is to avoid two problems that arise when fossil and recent taxa are combined in a single classification. The first is the proliferation of Linnaean ranks required to represent the levels of inclusiveness in such classifications. The second is the need to change the ranks of taxa in those classifications when newly discovered fossil taxa are added. Once again, both the post-Linnaean system and the annotated system deal with these problems, but the post-Linnaean system does so in a more consistent fashion. The annotated system attempts to resolve these problems by breaking taxa into two types: recent taxa that get Linnaean ranks, and fossil taxa that have no Linnaean ranks but are given the rank of plesion. The post-Linnaean system, on the other hand, does not posit a non-Linnaean rank, nor does it treat taxa differently. It simply does not assign ranks to any taxa. Consequently, the post-Linnaean system alleviates the problems associated with classifications of fossil and recent taxa in a simpler and more consistent manner.

A more cumbersome aspect of the annotated system is the assignment of "sedis mutabilis" to trichotomies and polytomies classified by phyletic sequencing (Conventions 3 and 5). Phyletic sequencing avoids the proliferation of categories by listing taxa on an asymmetrical clado-gram in the order of their branching, and by assigning those taxa the same rank (see the example in Section 6.3). But because phyletic sequencing assumes dichotomous branching, it cannot be used to accommodate trichotomies and polytomies. Convention 5 of the annotated system proposes a way of classifying trichotomies and polytomies by marking such taxa with the phrase "sedis mutabilis." Notice how one convention spawns another convention here. Phyletic sequencing is first introduced to avoid the proliferation of ranks. That creates the problem of how to deal with trichotomies and polytomies. The phrase "sedis mutabilis" is then introduced to handle those taxa that cannot be represented by phyletic sequencing. The post-Linnaean system cuts through these epicycles of rules by eliminating the use of Linnaean ranks. Both systems achieve the aim of not producing more Linnaean ranks. But once again, the post-Linnaean system does so in a simple and more consistent manner, whereas the annotated system requires an amendment upon an amendment to the original Linnaean system.

Let us consider another problem facing the annotated system, one that pits that system against cladism. Because the annotated system requires the assignment of mandatory ranks, a cladistic supporter of the Linnaean system is faced with two options when classifying ancestral taxa: either assign redundant ranks to a single monophyletic taxon, or posit the existence of a paraphyletic taxon. In the first option, the assignment of multiple Linnaean categories to the same taxon renders those categories empty designations. In the second option, the assignment of redundant categories is avoided by positing nonmonophyletic taxa (an ancestral taxon would be considered part of a higher taxon rather than equivalent to it). Either option is an unhappy one for a cladist who supports the Linnaean system. The post-Linnaean system offers the cladist a way out of this dilemma. No mandatory ranks are assigned to taxa, so there is no need to have either redundant categories or nonmonophyletic taxa. The post-Linnaean system can easily accommodate cladism while the annotated system cannot. The problem here for the annotated system is a fundamental one that cannot be easily rectified. Once mandatory Linnaean categories are assigned, either those categories become empty designations or nonmonophyletic taxa are allowed. Griffiths (1973, 1974) observed this some time ago and came to the conclusion that cladists need a non-Linnaean system. More recently, de Queiroz and Gauthier (1992) and Brummitt (1997a) have presented a similar line of reasoning.

The post-Linnaean system easily accommodates the needs of cladists, but at the same time it is not a strictly cladistic system. The post-Linnaean system is designed to be used by both cladists and evolutionary taxonomists. The annotated system, on the other hand, is firmly wedded to cladism. Consider Convention 6 of the annotated system. It allows the classification of nonmonophyletic taxa as long as such taxa are not assigned Linnaean ranks. Given this constraint, an evolutionary taxonomist would be hard-pressed to construct a Linnaean classification containing many paraphyletic taxa. The annotated system's inability to construct Linnaean classifications based on noncladistic principles goes against a commonly stated aim of taxonomy. Recall that the preamble of the *International Code of Zoological Nomenclature* (quoted in Section 6.5) explicitly states that the code's aim is to promote stable, universal, and unique taxon names, while not restricting "the freedom of taxonomic thought or action." The annotated Linnaean system violates that aim by ruling out Linnaean classifications based on the principles of evolutionary taxonomy. The post-Linnaean system is more accommodating. In fact, it has the virtue of addressing cladistic concerns, as in the case of ancestral taxa, while at the same time not forgoing the possibility of constructing noncladistic classifications. The post-Linnaean system has the practical advantage of being a tool for cladists and noncladists. Wiley's annotated system is firmly committed to cladism and only cladism.

Another virtue of the post-Linnaean system is its ability to resolve a number of problems that affect both cladistic and noncladistic taxonomy, such as the proliferation of categories and the existence of outdated rules of nomenclature. Consider the problems of nomenclature. In both the traditional and annotated systems, the ranks of most taxa are written in their names. This simple rule leads to unnecessary instability and confusion in biological taxonomy. Taxon names must be changed when taxa are reclassified, and different names must be assigned to a single taxon when biologists disagree over its rank or placement. These are just two problems caused by the Linnaean rules of nomenclature; many others exist (Section 6.4). As we have seen, the annotated system assumes the traditional system's rules of nomenclature without amendment, and thus the annotated system assumes all of the traditional system's nomenclatural problems. The post-Linnaean system cuts this Gordian Knot by separating the function of naming a taxon from that of indicating its taxonomic position. The hierarchical relation of a taxon in a post-Linnaean classification is indicated by either numerical prefix or indentation, not by its name. The forenames and suffixes of previously named taxa no longer indicate categorical rank; and newly discovered taxa are given forenames and suffixes that are rank neutral. Moreover, the post-Linnaean system overcomes the problems of traditional nomenclature without disturbing the names of previously named taxa. Consequently, stability is preserved in two ways: the alteration and multiplication of names due to revision and disagreement is halted; and no taxon names are altered in the switch to a post-Linnaean system.

The post-Linnaean system enhances biological nomenclature in another way. According to Wiley's Rule 2, "[t]he relationships of taxa within a classification must be expressed exactly" (1981, 200). Ax notes that this rule is "a necessary truism" (1987, 234). Wiley's recommendation here is a reasonable one. When a biologist constructs a classification, she provides a hypothesis concerning the relations among a group of taxa. Ideally, that classification should represent those relations in as clear and explicit a manner as possible. As a cladist, Wiley believes that cladistic classifications provide the clearest representations of taxonomic relations. A standard argument offered by cladists is that their classifications more clearly express hypothesized relations among taxa than classifications offered by evolutionary taxonomists (see Eldredge and Cracraft 1980, and Ridley 1986). I do not want to enter the debate between cladists and proponents of other schools of taxonomy. Instead, I would like to suggest that one way to enhance the clarity of classifications is to follow Recommendation 11 of the post-Linnaean system, namely, define taxon names with phylogenetic definitions.

Recall that de Queiroz and Gauthier's (1990, 1992, 1994) phylogenetic definitions explicitly cite the evolutionary relations that bind organisms and subtaxa into more inclusive taxa (see Section 7.3). The type method for defining taxon names, the one Wiley promotes in his 1981 text, defines taxon names by *implicitly* referring to those relations. Phylogenetic definitions make the appropriate phylogenetic relations explicit and in doing so further the aim of Wiley's Rule 2. Cladists are not the only taxonomists who can benefit from the use of phylogenetic definitions. As we saw in Section 7.3, de Queiroz and Gauthier suggest that such definitions can be used to define the names of nonmonophyletic taxa. For example, the phylogenetic definition of Reptilia might be "the most recent common ancestor of Mammalia and Aves and all descendents except Mammalia and Aves." So both cladists and evolutionary taxonomists can explicitly represent the relations of taxa by using phylogenetic definitions. Here, then, is another improvement the post-Linnaean system offers over the annotated system.

Let us step back and summarize the above comparison of the annotated and post-Linnaean systems. The post-Linnaean system drops the misleading and ontologically vacuous categories of the traditional system; the annotated system does not. The post-Linnaean system separates the function of naming and classifying taxa, thereby resolving a number of problems that stem from Linnaean nomenclature. The annotated system provides no changes in Linnaean nomenclature and is burdened with the naming problems of the traditional system. Both systems address the problems of redundant categories and the proliferation of ranks, but the post-Linnaean system does so in a simpler and more coherent manner. Finally, the annotated system does not resolve the conflict between the requirement of mandatory categories and desire for classifications containing only monophyletic taxa. The post-Linnaean system avoids that conflict and is thus a better system for cladists. Yet unlike the annotated system, the post-Linnaean system is not wedded to cladism. Thus the post-Linnaean system has the added benefit of being neutral among cladistic and evolutionary approaches to taxonomy.

In many areas, the post-Linnaean system is theoretically and pragmatically sounder than the annotated Linnaean system.[3] This is not to say that the post-Linnaean system is a perfect system. Like any system of nomenclature it has its problems. Many of those problems were addressed in Sections 7.1 and 7.2. Still, a pressing and overarching concern has not been discussed. Many authors question the practical merit of replacing the Linnaean system with a non-Linnaean alternative. We have seen a number of arguments for adopting the post-Linnaean system, but in the end does it make *practical sense* to throw out the Linnaean system? This concern is the topic of the next section.

8.3 THEORY AND PRACTICE

The acceptance of a scientific theory turns on a confluence of factors. Empirical adequacy rates high. A new theory should make the same successful predictions as its predecessors, and preferably more. After empirical adequacy come such theoretical virtues as simplicity, explanatory import, and coherence with other theories. Then there are practical concerns: How easy is a theory to use? Does its application require technology we lack? How convenient is it to abandon the current theory and adopt a new one?

When a new system of nomenclature is being considered, all these considerations are important as well. Systems of nomenclature provide methods for constructing representations of the world, namely, scientific classifications. As we saw in Chapters 2 and 5, biological systematists would like their classifications to be empirically adequate. Given that desire, classifications of the organic world should be vulnerable to empirical evidence, and they should also cohere with our best accepted empirical theories. Then there are practical concerns that affect the desirability of a system of nomenclature: Are its classifications easy to understand? Are the names in its classifications as stable as possible? Does it prescribe any outdated and meaningless tasks? A more general concern when adopting a new system is the difficulty of switching to a new set of nomenclatural rules.

This section considers whether it makes practical sense to replace the Linnaean system with the post-Linnaean system. An obvious concern when contemplating the adoption of the post-Linnaean system is whether the merits of the post-Linnaean system sufficiently outweigh the inconvenience of abandoning the Linnaean system. The aim of this section is not to provide a knockdown argument for adopting the post-Linnaean system, but to show that switching to the post-Linnaean system is not as impractical as it might seem at first. Despite initial appearances, adopting the post-Linnaean system, or some other suitable non-Linnaean system, might be the practical thing to do.

Let us begin by noting that the majority of reasons offered in this volume for replacing the Linnaean system are pragmatic, not theoretical. Furthermore, those pragmatic reasons are based on the sorts of desiderata biologists cite when they choose a system of nomenclature. Here is a sample of those criteria.

- A system of nomenclature should provide stable names and classifications (*International Code of Zoological Nomenclature*, Wiley 1981, de Queiroz and Gauthier 1992, Brummitt 1997b).
- It should assign unique names to taxa (*International Code of Zoological Nomenclature*, Wiley 1981, de Queiroz and Gauthier 1992).
- It should not prescribe meaningless procedures (Hennig 1969, Ax 1987).
- It should not be the cause of useless debates (Hennig 1969, Ax 1987).

- It should be a simple and general system (*International Code of Zoological Nomenclature*, Wiley 1981, de Queiroz and Gauthier 1992).
- Its classifications should clearly convey taxonomic information (Mayr 1969, Eldredge and Cracraft 1980, Wiley 1981).

With these pragmatic virtues, taxonomists also cite the theoretical virtue of providing classifications framed in the proper theoretical context (Eldredge and Cracraft 1980, de Queiroz and Gauthier 1992).[4] This latter virtue is what philosophers often refer to as "intra-theoretic consistency." Let us see how the Linnaean and the post-Linnaean system fare on these virtues. We will start with theoretical consistency, but spend most of our time with the pragmatic virtues.

As we saw in Section 6.5, the question of whether a system of nomenclature provides classifications within the proper theoretical context boils down to the question of whether its classifications are consistent with the theory of the day. Even this description, however, needs further refinement because the notion of "consistency" is itself ambiguous. Here I will use a fairly weak version of consistency. A system of nomenclature should not contradict the tenets of those theories that describe the entities being classified. For example, a system of nomenclature should not construct classifications that divide the different life stages of a single organism into different taxa – such a system of nomenclature would contradict accepted theories concerning the individuality of organisms. This weaker formulation of consistency derives from the desire to have systems of nomenclature that allow a degree of taxonomic freedom. More specifically, a system of nomenclature should not be overly wedded to a particular school of taxonomy. For instance, a system of nomenclature should not be a strictly cladistic system that only allows the classification of organisms according to cladistic principles. Still, some theoretical constraints should be placed on a system of nomenclature. A system of nomenclature should not construct classifications, nor contain tenets, that contradict current theory.

Many of the theoretical assumptions that underwrite Linnaeus's original system contradict contemporary evolutionary theory. Linnaeus's essentialist and creationist assumptions concerning the nature of taxa are at odds with current biological theory. Similarly, Linnaeus's motivation for using hierarchical classifications – that God created a world of hierarchically arranged organisms and taxa – has been rejected. Many of the assumptions that underwrite his rules of nomenclature and his sexual system for classifying plants suffer a similar fate. For example, Linnaeus posited the use of binomials as a way for biologists to memorize the positions of all the species in a kingdom. However, Linnaeus incorrectly assumed that God created a manageable set of taxa and that no new taxa would arise (Section 6.1).

A defender of the Linnaean system would, of course, agree that the vast majority of Linnaeus's theoretical assumptions are obsolete. Nevertheless, she could argue that we can continue to use various aspects of the Linnaean system without adopting his outdated assumptions. For instance, we can and do use his hierarchy while dropping the assumption that God created the biological hierarchy. Instead we assume that the hierarchy is the result of evolution. Similarly, we use Linnaeus's rules of nomenclature without adopting his motivations for those rules. And we can continue using the Linnaean categories without the assumption that taxa have essences or that those categories were created by God. In fact, the three aspects of the Linnaean system just highlighted – its categorical ranks, its rules of nomenclature, and the assumption that classifications should be hierarchical – are sufficiently theoretically neutral that they have withstood the change from creationism to evolutionary biology.

Still, the Linnaean system does carry unwanted theoretical baggage. There may be a consensus among biologists that taxa lack essences but the assumption that the Linnaean categories have essences is widely held. Taxa of a single category are often considered comparable, and taxa in different categories are considered importantly different. A major theme of this volume is that such assumptions no longer hold in light of evolutionary theory. Of course, a contemporary proponent of the Linnaean system might agree that the Linnaean categories lack essences but suggest that we should continue using those categorical ranks as indicators of subordination. This suggestion would render the Linnaean hierarchy theoretically innocuous. However, if the history of biology is any indication, we won't drop the past meanings of the Linnaean categories any time soon. One hundred years after the Darwinian revolution, the Linnaean ranks still carry essentialist meanings. The association between essentialism and the Linnaean categories is hard to break. Consider the fact that biologists and philosophers readily acknowledge that species taxa lack essences, but they have a much harder time allowing that the species category does not have an essence.

Notice, also, that if we keep the Linnaean categories but think that they are mere notational devices, then we have the awkward requirement of indicating meaningless ranks within the names of taxa. Species are given binomials even though there is no species category in nature. Similarly, the name of a higher taxon indicates that it belongs to particular categorical rank though no such category exists. So even if we were able to rid the Linnaean system of its problematic theoretical assumptions, those assumptions would continue to haunt us in the traditional rules of nomenclature.

Thus far I have suggested that the contemporary Linnaean system is not theoretically neutral but relies on outdated assumptions. The post-Linnaean system, on the other hand, does not carry the theoretical baggage of the traditional system. It is worth adding that the post-Linnaean system is more theoretically neutral than the annotated system. The post-Linnaean system is neutral when it comes to the principles of cladism and evolutionary taxonomy. The annotated system carries an inherent conflict between the Linnaean system and cladism. Redundant categories must be assigned to monobasic taxa, but that can be done only at a cost of allowing paraphyletic taxa or rejecting the Linnaean requirement of mandatory categories. The post-Linnaean system does not assign mandatory categories, so it avoids this conflict with cladism. The post-Linnaean system is also more accommodating to noncladistic classifications than the annotated system because paraphyletic taxa are not labeled as non-natural.

Perhaps some would object that the post-Linnaean system is too theoretically neutral, that a system of nomenclature should be more closely aligned with a particular school of taxonomy. However, I take the preamble of the *International Code of Zoological Nomenclature* seriously:

> The object of the Code is to promote stability and universality in the scientific names of animals, and to ensure that each name is unique and distinct. All its provisions are subservient to these ends, and none restricts the freedom of taxonomic thought or action (cited in Mayr 1969, 301).

To preserve freedom of taxonomic thought, a system of nomenclature should not be wedded to one school of taxonomy. Nevertheless, some theoretical constraints should be placed on a system of nomenclature. On the one hand, the post-Linnaean system better coheres with current theory than the annotated system – it does not posit theoretically suspect categories. On the other hand, the post-Linnaean system offers more taxonomic freedom than the annotated system by treating monophyletic and nonmonophyletic taxa equally. The post-Linnaean system achieves the precarious position of being more theoretically sensitive than the annotated system, yet allowing greater taxonomic freedom.

Let us now turn to the pragmatic virtues listed earlier. Biologists of all stripes would like the names of taxa to be stable and unique. As we have seen, both the traditional and annotated Linnaean systems score poorly on these virtues. In the Linnaean systems, the positions and ranks of taxa must be incorporated in their names. That requirement causes unwanted instability during revision because the names of taxa must be altered. The post-Linnaean system avoids such instability. The ranks and positions of taxa are not indicated by their names, so taxon names remain constant during revision. When it comes to the desire for unique names, the Linnaean system, again, rates poorly. If two biologists disagree on the position of a taxon, the Linnaean system forces them to assign different names to the very same taxon. This inconvenience is avoided in the post-Linnaean system because the functions of naming a taxon and indicating its taxonomic position are divorced. Two biologists can disagree on the placement of a taxon yet refer to that taxon with the same name.

The Linnaean rules of nomenclature are not the only area of the Linnaean system that is pragmatically flawed. The assignment of categorical ranks to taxa also causes unnecessary inconvenience. As we have seen, the higher Linnaean ranks do not reflect true divisions in nature. Moreover, various suggestions for correlating the taxa of a particular higher rank have come up empty-handed. All orders, for example, do not arise during the same geological period, nor do they contain similar numbers of subordinate taxa (Section 6.3). If the higher Linnaean categories are heterogeneous collections of taxa, then biologists should be freed from the theoretically empty task of having to assign higher Linnaean ranks to taxa. The same holds for the assignment of taxa to the species category (Chapter 4). Those taxa typically thought of as species are noncomparable entities: some are groups of organisms connected by interbreeding, others are lineages of asexual organisms; some are monophyletic, others are paraphyletic. It is doubtful that a feature exists that distinguishes species taxa from all other types of taxa. Biologists should be freed from the theoretically suspect task of having to determine whether a taxon is a species or another type of taxon.

Hennig (1969, xviii) goes further and argues that the assignment of

Linnaean ranks is not only a meaningless task but one that leads to needless disagreements among biologists. The Linnaean system requires that each taxon be assigned a Linnaean rank (or the rank of plesion in the annotated system). However, biologists will at times disagree on the rank of a taxon. According to Hennig, such disagreements lead to "unfruitful debates" among taxonomists. Another unfruitful debate spawned by the need to assign ranks to taxa is the debate over the correct definition of the species category. If the species category does not exist, then much time and effort has been wasted in trying to define that category. In summary, the requirement of assigning ranks to taxa causes biologists to perform a number of inconvenient activities: the assignment of ranks to taxa itself; the incorporation of ranks in the names of taxa; debates over the ranks of individual taxa; and debates over the meanings of ranks. Each of these activities is a time-consuming distraction that no longer has theoretical merit. Biologists can avoid these activities by switching to a system, like the post-Linnaean system, that does not assign any ranks to taxa.

One other pragmatic virtue, promoted by both supporters and detractors of the Linnaean system, is that systems of nomenclature be simple and general. Here the motivation is not aesthetic, nor is it the epistemological idea that simpler hypotheses are more likely to be true. Here the concern is firmly practical. The simpler the system of nomenclature, the fewer rules there will be to learn and remember. In order for a system of nomenclature to be simple, it must contain rules that apply to a wide range of taxa. If some rules apply to some taxa, other rules to different taxa, and still other rules to a third set of taxa, a system's nomenclatural rules become piecemeal and not general. Such complexity, or lack of coherence, makes a system harder to apply, for different rules must be used in different cases, and workers must learn which instances are special cases.

The traditional Linnaean system is quite general and consistent in applying ranks to taxa – all taxa are assigned Linnaean ranks. The assignment of ranks in the annotated Linnaean system is more piecemeal. Some taxa are given Linnaean ranks that indicate their hierarchical relations, others are not. Those taxa that lack Linnaean ranks indicating their levels of inclusiveness lack such ranks for different reasons. Fossil taxa, for example, are not assigned Linnaean ranks but are given the rank of plesion. Taxa in a phyletic sequence are assigned a single Linnaean rank rather than different ranks indicating their distinct levels of inclusiveness. Then there are trichotomies and polytomies within phyletic sequences that are given the tag "sedis muta-bilis" to indicate that they are not the result of asymmetric branching. All other taxa are assigned Linnaean ranks reflecting their hierarchi-cal relations.

Individually, the rules of the annotated system have limited appli-cation. Collectively, they form an awkward set of rules cobbled together to save the Linnaean system from various problems. De Queiroz and Gauthier (1992, 450) are right when they remark that the annotated system is an "ad hoc modification of a body of conventions based on a pre-Darwinian world view." The post-Linnaean system offers a simple set of rules that individually have large scope. At the same time, the rules of the post-Linnaean system overcome the prob-lems facing the Linnaean system. The relevant point here, however, is ease of use. The annotated Linnaean system consists of a cumbersome and piecemeal set of rules, whereas the post-Linnaean system contains a smaller set of more generally applicable rules.

We have seen a number of reasons why the Linnaean system is impractical. Its rules of nomenclature cause instability and confusion over taxon names. Its requirement that taxa be assigned ranks is a source of fruitless debates. Furthermore, the annotated Linnaean system's conventions for assigning ranks are cumbersome and ad hoc. In contrast, the post-Linnaean system overcomes these practical prob-lems and is easier to use. Recall the primary question of this section: Does it make practical sense to switch from the Linnaean system to the post-Linnaean system? The answer so far is yes. The post-Linnaean system's rules are more convenient to use, and adopting them would make the job of biological taxonomists easier.

However, there are practical concerns that have not yet been dis-cussed. We have seen from the perspective of biology that there are ample pragmatic reasons for adopting the post-Linnaean system. But we have not considered any concerns that might be raised from outside of biology. One might wonder from a political and social perspective whether it is wise to eliminate the Linnaean system. Hull (personal communication), for example, asks if the acceptance of a non-Linnaean system might discourage governments from taking an active role in environmental protection. Hull's concern is twofold. First, current envi-ronmental laws are often formulated in terms of species. If we reject the Linnaean hierarchy and the species category, then many environ-mental laws will become useless. Second, if some politicians discover that biologists are divided on the proper way to construct classifications, they might argue that legislating environmental protection is inappropriate because scientists disagree on what to protect. Imagine a case in which a supporter of a phylogenetic species concept recommends protecting a species of asexual organisms, while a supporter of the Biological Species Concept counters that there is no species here to protect.

This latter fear, however, overplays how much controversy exists in biological taxonomy. Undoubtedly controversy exists over the basal units of biological classification: some biologists are species monists, others are species pluralists; some believe that asexual organisms form basal units, others do not. Despite these disagreements, there is some agreement concerning the importance of biospecies. Critics of the Biological Species Concept (BSC) argue that interbreeding approaches to species neglect much of life on this planet, namely, all asexual organisms. In addition, they question the effectiveness of gene flow within populations of interbreeding organisms. Nevertheless, both promoters and detractors of the BSC concur that those taxa that are maintained by the interactive process of interbreeding are significant taxonomic units (Sections 2.5, 4.1, and 4.2). When interbreeding is effective, it does cause the existence of significant basal units. Promoters and detractors of the BSC, of course, disagree on the number of such taxa, but true biospecies, I would suggest, are uncontroversial candidates for recognition and protection.

Matters may seem more controversial when considering groups of asexual organisms; however, they need not be controversial. Although proponents of the Biological Species Concept deny that groups of asexual organisms form species, they can allow that some groups of asexual organisms are worthy of protection. In other words, biologists can agree on which groups of organisms should be protected even though they disagree over the Linnaean ranks of those groups. For instance, a proponent of a phylogenetic species concept may consider a group of asexual organisms a species, whereas a supporter of the BSC would deny that they form a species; nevertheless, those biologists can agree that the group in question should be protected. This is a case where dropping the Linnaean ranks would allow biologists to speak with a more unified voice: they can say which taxa should be protected without arguing over the Linnaean status of those taxa.[5]

Let us return to Hull's first worry. Environmental laws often refer to species, so if we eject the term "species" from biological taxonomy, then our environmental laws will become empty. We can extend this

288

worry further. If we eject the term "species" from biological taxonomy, then almost all biological textbooks and field guides will become obsolete. Here, then, seems to be a formidable practical problem with replacing the Linnaean system with a post-Linnaean system: the term "species" is so firmly entrenched in our laws and texts, that we cannot realistically expect to eliminate it, except, perhaps, in limited technical settings.

Again, the problem sounds worse than it is. We could, for example, adopt the post-Linnaean system with the proviso that the term "species" be retained for pragmatic reasons. According to Beatty (1985), this was Darwin's tactic when he advanced the theory of evolution. Darwin was skeptical of the existence of the species category, yet in order to communicate with other biologists he used the term. "Species" was not a completely empty term for Darwin, for he did believe in the existence of those lineages commonly called species. He just doubted that an appropriate definition of species could be given within evolutionary theory.

Contemporary biologists could adopt a similar strategy. They could recognize that the species category is heterogeneous and does not exist. At the same time they could allow the continued use of "species" in our laws and books. Ideally, the reference to species should be eliminated, and in more technical discussions that would be recognized. In those technical works the term "species" would be disambiguated. Biologists would explicitly state whether the basal units they are talking about are biospecies, phylospecies, or ecospecies. Nevertheless, in less technical situations, the ambiguous and misleading term "species" would remain. It is worth recognizing that progress in biological taxonomy does not require that we immediately reject all elements of the Linnaean system. If for pragmatic reasons the term "species" needs to be retained, then we should keep it, so long as biologists recognize that the species category is at best a nominal concept.

Ardent critics of the Linnaean system may be unhappy with this suggestion and counter that the Linnaean categories, including the species category, should be relegated to the history of science. With scientific progress often comes a change in terminology, both in and outside science. Biological taxonomy and the Linnaean categories are no different. If the terms of the Linnaean hierarchy are to be rejected in biological taxonomy, then we should bring our laws and textbooks up to date. I agree with this argument up to a point. Biologists who think the

Linnaean hierarchy should be abandoned should work to that end. Yet while doing so they still need to communicate with other biologists, as well as legislators and the general public. I do not see how detractors of the Linnaean hierarchy can completely avoid using the term "species." Post-Linnaean taxonomists would be wise to consider Darwin's strategy for employing the term.

Finally, suppose the post-Linnaean system is a superior system, that it is theoretically sounder, and that its adoption would make the job of biological taxonomists easier. One might argue that despite these virtues, there is little hope of replacing the Linnaean system given its widespread use both in and outside biology. Allan Larson (personal communication) offers an analogy with the continued use of the QWERTY keyboard. The upper left-hand keys of most English keyboards have the arrangement QWERTY. This arrangement of keys is suboptimal for typing; other arrangements would make for more efficient and quicker typing. Nevertheless, the typewriter (and computer) keyboard is "so burdened with historical usage" that we do not adopt a better keyboard arrangement. Typewriter keyboards, in other words, are stuck in the grip of historical contingency. So too are the rules of biological nomenclature. The practices that govern the construction of biological classifications are locked in a form of social stasis.

The post-Linnaean system acknowledges that stasis in practice can be very hard to overcome, and in some instances not worth fighting. Thus Recommendations 3, 7, and 9 suggest that the names of previously identified taxa be maintained. Not only is it too much bother to relearn the names of most taxa, but the vast majority of biologists would simply refuse to do so. Stasis in practice places a strong constraint on how far a system of representation can be altered. The post-Linnaean system attempts to be modest in light of such stasis. Nevertheless, the post-Linnaean system is a radical departure from long-standing tradition, for it recommends abandoning the Linnaean ranks and various rules of nomenclature.

Given the radical nature of these changes, one might wonder whether it would be too hard to replace the Linnaean system. We should not be too pessimistic. The history of science is full of episodes in which a generally accepted empirical theory or system of classification has been replaced by a radically new one. The same may occur in biological nomenclature. We should not foreclose that possibility at the start. Moreover, the notion of scientific change should not be foreclosed merely because it is change; otherwise, no scientific progress would occur. Promoters of non-Linnaean systems need to counter the entrenchment of the Linnaean system by developing non-Linnaean alternatives. They need to show that those alternatives are more efficient, and they need to broadcast those results. Many authors recognize the problems of the Linnaean system but counter that we must live with that system because "it is the only generally accepted system we have" (Jeffrey 1989, 14). Part III of this volume illustrates the problems of the Linnaean system. More important, it shows that viable alternatives exist and that replacing the Linnaean system is well worth the effort.

Notes

INTRODUCTION

1. It is worth emphasizing that the suspicion voiced here concerns the existence of the Linnaean categories and *not* the reality of taxa. We can remain confident in the reality of *Homo sapiens* and *Homo*, and that *Homo sapiens* is a part of *Homo*. The problem is that the designations of "species" and "genus" may add nothing to these facts.
2. Some taxonomists will take issue with the claim that there are four major schools of taxonomy, arguing that one or two of the mentioned schools are not viable options. Proponents of rejected schools, of course, would disagree. Putting such disagreements to one side, no taxonomist would deny that there is more than one major school of biological taxonomy.

CHAPTER 1

1. For further problems with essentialism see Dupré (1981, 1986, 1993), Mellor (1977), and DeSousa (1984).
2. We will return to the application of Boyd's homeostatic cluster account to biological species in Chapter 3.
3. The qualifier "at the same time" refers to the co-existence of an entity's spatial parts, not the causal processes that connect those parts. The following discussion respects the common philosophical assumption that a cause and its effect cannot occur at the same time.
4. This seems to be what is happening in the *Ship of Theseus* problem (a standard philosophical conundrum). Suppose we replace the pieces of the ship of Theseus as they wear out. Over time the ship is completely made up of replacement parts. Suppose, also, that all of the original parts have been collected and assembled into a ship. Which ship is the ship of Theseus: the ship with the replacement parts or the one with the original parts? If the original parts were merely discarded we might be inclined to think that the ship with the replacement parts is the ship of Theseus. But once we are told that the original parts might be parts of another ship, determining which ship is the ship of Theseus becomes harder.

5. In Chapter 6 we will return to the question of whether speciation due to hybridization occurs and whether it is inconsistent with the desire for hierarchical classifications. As we shall see, Hull (1964) argues that such speciation events do occur in nature and that they are problematic for the Linnaean hierarchy. Wiley (1979), on the other hand, thinks that such speciation events can be accommodated by his annotated Linnaean hierarchy.

CHAPTER 2

1. For a full introduction to the modern synthesis, see Provine (1971), and Mayr and Provine (1980).
2. The definition of monophyletic taxon used here is one proposed by cladists. Evolutionary taxonomists offer another definition of monophyly that captures both monophyly (as defined by cladists) and paraphyly (see Simpson 1961 and Mayr 1969). I have opted for the cladistic definition because it makes clear the distinction between monophyly (as defined by cladists) and paraphyly.
3. As mentioned earlier, outgroup comparison is just one of several methods cladists use for distinguishing synapomorphies from symplesiomorphies. Two other methods are the embryological criterion and the paleontological criterion. I will not discuss these here. Ridley (1986, 66–9) and Ridley (1993, 462–4) contain introductory discussions of these criteria.
4. Ironically, Darwin does not provide a positive description of species in the *Origin*. In fact, he seems doubtful whether the species category exists (see Darwin 1859[1964], 51, 52, 485). We will discuss the reality of species and Darwin's view on the matter in Chapters 4 and 6.
5. At this juncture, we are not interested in the sorts of definitions biologists provide for the names of species taxa. We will turn to that issue in Chapter 6.
6. This makes sense given that the two major proponents of the biological species concept, Mayr and Dobzhansky, were instrumental in applying the insights of the synthetic theory of evolution to biological taxonomy. The biological species concept is one of those applications.
7. I have divided phylogenetic species concepts into those that are process oriented and those that are pattern oriented. Baum and Donoghue (1995) divide phylogenetic species concepts into history-based approaches and character-based approaches. My division lines up with theirs. Davis (1995) offers an even more fine-grained division of phylogenetic species concepts.
8. Mishler and Brandon (and Mishler and Donoghue 1982) provide other reasons for preferring a phylogenetic species concept over the biological species concept. One reason is that asexual organisms form species, yet such organisms cannot form species according to the biological species concept. Another reason is their doubt over the effectiveness of interbreeding in maintaining the existence of species.
9. Whether species can form monophyletic taxa is a controversial issue (see, for example, de Queiroz and Donoghue 1988). We will turn to this issue in later chapters, especially in Chapters 6 and 8.

CHAPTER 3

1. Ereshefsky 1988 critically reviews the claim that the individuality of species has implications for macroevolutionary theory.
2. The first sentence in this quotation should not lead one to think that Mayr is endorsing species essentialism. Mayr is a vigorous opponent of species essentialism. Furthermore, his account of isolating mechanisms is at odds with essentialism. No particular isolating mechanism is *necessary* for species membership since the breakdown of an isolating mechanism does not mark a speciation event (Mayr 1963, Chapter 6). Moreover, no particular isolating mechanism is *necessary* for membership in a species because conspecific organisms can vary in their isolating mechanisms (Mayr 1963, 106–7).
3. Griffiths (1997) more carefully applies Boyd's homeostatic cluster approach to biological taxa. Griffiths explicitly recognizes the historical basis for picking certain homeostatic mechanisms as mechanisms of a particular taxon. But again, the glue that binds organisms and their homeostatic mechanisms is genealogy, not the qualitative features of taxa. The historical approach, not a qualitative approach like Boyd's, is doing the crucial work here.
4. I have used a description of local populations that allows the existence of populations of asexual organisms. Supporters of interbreeding species concepts, however, limit local populations to only populations of sexual organisms (see for example, Mayr 1963, 136).
5. This statement does not rule out the occurrence of similar traits in different species due to separate evolutionary events. That is, it does not rule out instances of parallel evolution that result in analogous traits. The argument merely shows that any talk of similar traits from a common source requires genealogical connections.
6. It is worth noting that the distinction between asexual species and sexual species is not an absolute one (Templeton 1998[1992], 164–5). Barley and blue-green algae, for example, are considered asexual species. The vast majority of their members are self-mating. Nevertheless, there is occasional intragenerational gene exchange within these species. Such gene exchange, however, is extremely rare and insufficient to cause the unity of these species (Templeton 1998[1992], 164–5; Hull 1988, 429; also see Chapter 4, n. 1). Other species may consist of a more balanced mix of self-mating organisms and those that interbreed. It would be wrong to characterize such species as either asexual or sexual. The question of whether interbreeding binds such species into evolutionary units is an open one.
7. Lange (1995) counters that generalizations about particular species can be laws of nature despite their limited temporal scope. Following Dirac, Lange points out that because the universe expands and contracts, the constants of physical laws change such that physical laws hold for only a particular "cosmic epoch" (448). That may very well be the case. But notice that Lange and Dirac are talking about the stability of physical laws across great lengths of time – for instance, the time it takes for the universe to expand. The lengths of time that generalizations concerning particular species are stable is much,

much shorter. Lange's claim concerning physical laws and cosmic epochs has intuitive appeal, but when applied to generalizations concerning particular species, as Lange does, the notion of natural law is stretched to the breaking point.

CHAPTER 4

1. One objection against the existence of asexual species I have not considered is that lineages containing *only* asexual organisms rarely occur (Mayr 1963, 410ff.; Eldredge 1985, 200–1). That may be, but such lineages do not then fall under the interbreeding approach to species, as the objection implies. Consider Mayr's (1963, 410ff.) example of blue-green algae. Although genetic exchange occasionally occurs, it is not through meiosis. Yet meiosis is the process of sexual reproduction that distinguishes sexual from asexual organisms. Furthermore, the frequency of gene exchange in blue-green algae is very rare: one cell in 240,000 by one estimate, and one in 2 million by another (Hull 1980, 429). That rate is too low to render blue-green algae a cohesive gene pool of the sort required by the interbreeding approach (Hull 1987, 179).

2. If one does not like the example of asexual organisms, then consider cases of sexual species consisting of geographically isolated but stable populations. Such species are not maintained by gene flow between their populations but by other evolutionary forces. Templeton (1989[1992], 165) and others suggest that such cases are far from the exception among sexual species. Here, then, is another example of the widespread discrepancy between the interbreeding approach and other approaches.

3. As an example, consider a monophyletic interbreeding taxon that contains multiple monophyletic ecological taxa. Syngameons (discussed in Section 4.1) may be examples of this kind.

4. Kitcher's description of the phylogenetic approach is not quite right. Wiley (1981) allows other principles of division than reproductive isolation. So do other proponents of the phylogenetic approach (see Mishler and Donoghue 1982; Mishler and Brandon 1987; and Ridley 1989). Furthermore, Kitcher has incorrectly placed pattern cladist species concepts under the category of historical approaches. As we saw in Section 2.4, pattern cladists are not committed to species taxa, or any taxa, being genealogical entities.

5. Kitcher does not list phenetics as a legitimate structural species concept. Kitcher does not explain this omission, but I assume it is because phenetics does not attempt to capture theoretically significant similarities (see Section 2.2) whereas Kitcher's structural concepts do.

6. For additional problems with Stanford's form of species pluralism, see Ereshefsky 1998.

7. Still, it is not hard to envision empirical circumstances in which a single organism belongs to two different interbreeding species. Suppose an organism successfully interbreeds and produces fertile offspring with members in two species. Resultant offspring return to the other parent's species and successfully reproduce. The two species remain intact even though an occasional organism is a member of both gene pools.

8. Stanford (1995, 77–80) suggests a way of fleshing out Kitcher's criterion. According to Stanford, scientific theories provide explanatory schemata, and classifications render questions more tractable within a schema by highlighting divisions relevant to that schema. Species concepts that allow the construction of explanatorily useful classifications should be accepted, provided they are "neither redundant, boring, nor wrongheaded" (1995, 80). Stanford's formulation is more elaborate than Kitcher's. Yet it does not significantly depart from Kitcher's requirement that we should accept only those concepts that stem from currently successful theory (see Section 4.3 and Ereshefsky 1998).

CHAPTER 5

1. Rosenberg (1985b, 1990) also offers a naturalistic approach to methodology. See Ereshefsky (1995a) for a comparison of Laudan's and Rosenberg's approaches.
2. I thank John Dupré and Bradley Wilson for raising this point.
3. The vagueness of a general aim could spell trouble for an application of normative naturalism if that aim is so vague that no specific methodological rules can be derived from it. (I thank Elliott Sober for raising this concern.) In the next section, the suggested aim of biological taxonomy will be used to derive methodological rules for that discipline. As we shall see, specific methodological rules can be derived from that aim, despite its general nature.
4. Although the historical route is not employed here, it can be useful in analyzing the methodological rules of biological taxonomy. Mayr (1982, 174, 179, 191) and Atran (1990, 108), for instance, offer a historical case against the method of logical division. According to that method, entities are sorted into a hierarchy of classes such that each class is subdivided into two lower classes by a set of differentiating properties. As Mayr and Atran observe, the method of logical division breaks up natural groups (see Section 1.1). Thus that method fails to provide empirically accurate classifications that serve as a basis for making inferences about the organic world.
5. See Jeffrey 1983 for a general introduction. For an analysis of how empirical evidence affects hypotheses concerning phylogeny, see Sober 1988.
6. One may protest that the standards of intertheoretic and intratheoretic coherence are too conservative. Consider upstart taxonomic approaches that conflict with well-established theory but are not yet empirically tested. Don't the above standards too quickly nip in the bud fledgling taxonomic approaches that are perhaps worthy of further study? This is not a given according to the process of evaluation being recommended here. As I suggest in the next paragraph, the above standards are not individually necessary.
7. I have discussed vagueness concerning the number of primary rules a taxonomic approach should satisfy. Vagueness can also occur concerning the degree to which a taxonomic approach satisfies a particular rule. Again, such vagueness is far from fatal as long as that rule is useful in determining the fate of some taxonomic approaches. We will turn to an example of this latter form of vagueness in the next section.

8. For an introduction to the phenetic species concept and the other species concepts analyzed in this section see Section 2.5.

9. Recall that both pattern cladists and process cladists refer to their species concepts as "phylogenetic species concepts" (Section 2.5). We will examine a process cladist species concept shortly.

10. At one point, Mayr offered a version of the biological species concept that allows asexual organisms to form species (1982, 273). Since then he has backed away from the suggestion that asexual organisms form species (1987, 165–6).

11. Ghiselin (1987, 138) and Hull (1987, 179–80) suggest that we can deal with this conflict by relaxing the assumption that all organisms belong to species. But this suggestion does not resolve the problem outlined above because it assumes that exceptions to the biological species concept are minimal. Consider Hull's analogy between classification and a library. He writes that just as some library books may fail to be shelved, some organisms may not belong to species. This analogy appeals to our intuitions that a small percentage of speciesless organisms should not bother us. However, I do not think that Ghiselin and Hull are willing to allow that most organisms are not members of any species.

12. Some cladists attempt to circumvent this problem by asserting that though ancestral species are not monophyletic they are nevertheless real because they are individuals (Eldredge and Cracraft 1980, 80). This proposal, and more generally the problem ancestral species pose for cladism, will be discussed in Chapters 6 and 8.

CHAPTER 6

1. Here I follow descriptions provided by Cain (1958, 1959a), Larson (1971, 83–4) and Atran (1990, 174–5).

2. The following account is described in Larson 1971, 96ff.

3. The following account is found in Larson 1971, 100ff. and Hull 1985, 47ff.

4. Wiley (1981, 225ff.) also suggests altering the Linnaean hierarchy so that it can indicate the ancestry of hybrid taxa. That suggestion is similar to his other concerning stem species. Using Wiley's example, suppose the genus *Hus* is the result of hybridization between *Sus aus* and *Tus bus*. Then the line containing *Hus* in a classification should be written: "Genus *Hus* (*Sus aus* X *Tus bus*)."

5. The observation that the Linnaean system's naming rules cause unneeded instability in biological classification is made by many authors (for example, Michener 1964, 183ff.; Mayr 1969, 334; Brothers 1983, 36ff.; Ereshefsky 1994, 196; de Queiroz and Gauthier 1994, 2).

CHAPTER 7

1. One might worry that by using the terminology of the Linnaean system we run the risk of importing the Linnaean ranks into a post-Linnaean system. That is not the case. In the earlier discussion, the terminology of the Lin-

naean system is merely used to address that system's rules of nomenclature. In post-Linnaean classification, taxa merely occur at different levels of inclusiveness within particular classifications; no ranks, Linnaean or otherwise, are assigned.

2. The discussion mentions "newly discovered basal taxa," but what are basal taxa in a post-Linnaean system? In the Linnaean system they are species. In a post-Linnaean system, basal taxa would be the least inclusive genealogical entities recognized by a legitimate taxonomic approach. This answer brings us back to species pluralism (Chapter 4). Various approaches to species – phylogenetic, interbreeding, and ecological – highlight different basal lineages. For the interbreeding approach, basal taxa are reproductive groups bound by interbreeding and separated by reproductive isolation. For the phylogenetic approach, basal taxa are basal monophyletic taxa bound by one or more unifying processes. Different taxonomic schools highlight different basal taxa, and they do so because they highlight different genealogical chunks on the tree of life.

3. Type definitions should not be confused with either the descriptions or diagnoses that accompany definitions (Hull 1978[1992], 308). A description is a lengthy characterization of a taxon, highlighting which traits distinguish its members from those of closely related taxa. Diagnoses are short summaries of descriptions. Neither descriptions nor diagnoses are definitions; they are merely heuristic devices for identifying taxa. Moreover, they can fail as heuristic devices given the diversity within taxa.

4. I thank Elliott Sober (personal communication) for bringing this point to my attention.

5. Their proposal has received much attention. See Bryant 1994, 1996; Ghiselin 1995; Graybeal 1995; and Harlin and Sunberg 1998.

6. Bryant (1994, 1996) presents a number of problems with phylogenetic definitions. The vast majority of these problems, as he explains, can be avoided when the rules governing phylogenetic definitions are suitably amended. Bryant's intention is to develop, not criticize, the phylogenetic system. The following discussion of the phylogenetic system is not intended as a criticism but to raise a more general point about our inability to provide complete definitions of taxon names.

CHAPTER 8

1. A noncladist defender of the Linnaean system might go further and assert that the problem here is not with the Linnaean system, annotated or otherwise, but with cladism. Simply drop cladism and the Linnaean system will be preserved. However, this is not a viable option for at least two reasons. First, many of the problems facing the Linnaean system apply whether or not one is a cladist – for example, various pragmatic problems that arise from Linnaean nomenclature. Second, the advances of phylogenetic systematics cannot be swept under the rug. Any contemporary system of nomenclature worth its weight must accommodate phylogenetic taxonomy; otherwise, that system would be irrelevant to much of contemporary classification.

Notes

2. Notice that Wiley's decision to treat species, genera, families, orders, and phyla differently from other types of taxa stems from the long-standing use of the mandatory categories rather than anything intrinsically different about the taxa in those categories. This move nicely illustrates the ontologically vacuous nature of the Linnaean categories: categories are treated differently because of tradition rather than any empirical facts about them.

3. The post-Linnaean system addresses most of the situations that the annotated system is designed to cover, but it does not address them all. Specifically, Conventions 5 and 6 of the annotated system offer a way of classifying taxa with uncertain relations. Convention 11 proposes a method for classifying hybrid taxa while preserving the hierarchical nature of classification. The post-Linnaean system does not address these issues; nevertheless, nothing prevents recommendations like Wiley's Conventions 5, 6, and 11 from being incorporated into the post-Linnaean system. The post-Linnaean system is not complete, but needs to grow and evolve.

4. Chapter 6 contains a detailed discussion of these virtues (pragmatic and theoretical), as well as references to their use in the literature on biological systematics.

5 While I have highlighted areas where systematists agree, it is also worth noting that the existence of disagreement among systematists does not undermine the credibility of their discipline. Disagreement among scientists is a sign of a healthy discipline. Of course, too much internal strife can cripple a discipline, but that is far from the case in biological systematics.

References

Ackrill, J. (1981). *Aristotle the Philosopher*. Oxford University Press, Oxford.

Andersson, L. (1990). "The Driving Force: Species Concepts and Ecology," *Taxon* 39:375–82.

Aristotle (1984). *Parts of Animals*, in J. Barnes (ed.), *The Complete Works of Aristotle*, Volume 1. Princeton University Press, Princeton, pp. 994–1086.

Atran, S. (1990). *Cognitive Foundations of Natural History: Towards an Anthropology of Science*. Cambridge University Press, Cambridge.

Ax, P. (1987). *The Phylogenetic System*. Wiley and Sons, New York.

Ayers, M. (1981). "Locke versus Aristotle on Natural Kinds," *Journal of Philosophy* 78:247–72.

Barnes, D. and D. Bloor (1982). "Relativism, Rationalism, and the Sociology of Knowledge," in M. Hollis and S. Lukes (eds.), *Rationality and Relativism*. MIT Press, Cambridge, Massachusetts, pp. 21–47.

Baum, D. and M. Donoghue (1995). "Choosing among Alternative 'Phylogenetic' Species Concepts," *Systematic Biology* 20:560–73.

Baum, D. and K. Shaw (1995). "Genealogical Perspectives on the Species Problem," in P. Hoch and A. Stephenson (eds.), *Experimental and Molecular Approaches to Plant BioSystematics*. Missouri Botanical Garden, St. Louis, pp. 289–303.

Beatty, J. (1980). "Optimal-Design Models and the Strategy of Model Building in Evolutionary Biology," *Philosophy of Science* 47:532–61.

Beatty, J. (1981). "What's Wrong with the Received View of Evolutionary Theory," in P. Asquith and R. Giere (eds.), *PSA 1980*, Volume 2. Philosophy of Science Association, East Lansing, Michigan, pp. 397–426.

Beatty, J. (1982). "Classes and Cladists," *Systematic Zoology* 31:25–34.

Beatty, J. (1985). "Speaking of Species: Darwin's Strategy," in D. Kohn (ed.), *The Darwinian Heritage*. Princeton University Press, Princeton, New Jersey, pp. 265–81.

Beckner, M. (1959). *The Biological Way of Thought*. Columbia University Press, New York.

Boyd, R. (1990). "Realism, Approximate Truth, and Philosophical Method," in W. Savage (ed.), *Minnesota Studies in the Philosophy of Science XIV*. University of Minnesota Press, Minneapolis, pp. 355–91.

References

Boyd, R. (1991). "Realism, Anti-foundationalism, and the Enthusiasm for Natural Kinds," *Philosophical Studies* 61:127–48.

Briggs, D. and S. Walters (1984). *Plant Variation and Evolution*. Cambridge University Press, Cambridge.

Brothers, D. (1983). "Nomenclature at the Ordinal and Higher Levels," *Systematic Zoology* 32:34–42.

Brummitt, R. (1997a). "Taxonomy versus Cladonomy, a Fundamental Controversy in Biological Systematics," *Taxon* 46:723–34.

Brummitt, R. (1997b). "The Biocode Is Unnecessary and Unwanted," *Systematic Botany* 22:182–6.

Bryant, H. (1994). "Comments on the Phylogenetic Definition of Taxon Names and Conventions Regarding the Naming of Crown Clades," *Systematic Biology* 43:124–30.

Bryant, H. (1996). "Explicitness, Stability, and Universality in the Phylogenetic Definition and Usage of Taxon Names: A Case Study of the Phylogenetic Taxonomy of the Carnivora (Mammalia)," *Systematic Biology* 45:174–89.

Cain, A. (1958). "Logic and Memory in Linnaeus's System of Taxonomy," *Proceedings of the Linnaean Society of London* 169:144–63.

Cain, A. (1959a). "Taxonomic Concepts," *Ibis* 101:302–18.

Cain, A. (1959b). "The Post-Linnaean Development of Taxonomy," *Proceedings of the Linnaean Society of London* 170:234–44.

Cain, A. (1960). *Animal Species and Their Evolution*. Harper and Brothers, New York.

Carson, H. (1957). "The Species as a Field for Genetic Recombination," in E. Mayr (ed.), *The Species Problem*. American Association for the Advancement of Science, Washington, D.C., pp. 23–38.

Carson, H. and A. Templeton (1984). "Genetic Revolution in Relation to Speciation Phenomena: The Founding of New Populations," *Annual Review of Ecology and Systematics* 15:97–131.

Cartwright, N. (1983). *How the Laws of Physics Lie*. Oxford University Press, Oxford.

Colless, D. (1967). "An Examination of Certain Concepts in Phenetic Taxonomy," *Systematic Zoology* 16:6–27.

Cracraft, J. (1983[1992]). "Species Concepts and Speciation Analysis," in R. Johnston (ed.), *Current Ornithology*. New York: Plenum Press, pp. 159–87. Reprinted in M. Ereshefsky (ed.), *The Units of Evolution*. MIT Press, Cambridge, Massachusetts, pp. 93–120.

Cracraft, J. (1987). "Species Concepts and the Ontology of Evolution," *Biology and Philosophy* 2:329–46.

Crowson, R. (1970). *Classification and Biology*. Atherton Press, New York.

Darwin, C. (1859[1964]). *On the Origin of Species: A Facsimile of the First Edition*. Harvard University Press, Cambridge, Massachusetts.

Darwin, F., ed. (1877). *The Life and Letters of Charles Darwin, including an Autobiographical Chapter*. John Murray, London.

Davis, J. (1995). "Species Concepts and Phylogenetic Analysis – Introduction," *Systematic Biology* 20:555–9.

References

Davis, J. and K. Nixon (1992). "Populations, Genetic Variation, and the Delimination of Phylogenetic Species," *Systematic Biology* 41:421–35.

de Queiroz, K. (1988). "Systematics and the Darwinian Revolution," *Philosophy of Science* 55:238–59.

de Queiroz, K. (1992). "Phylogenetic Definitions and Taxonomic Philosophy," *Biology and Philosophy* 7:295–313.

de Queiroz, K. (1994). "Replacement of an Essentialistic Perspective on Taxonomic Definitions as Exemplified by the Definition of 'Mammalia'," *Systematic Biology* 43:497–510.

de Queiroz, K. (1997). "The Linnaean Hierarchy and the Evolutionization of Taxonomy, with Emphasis on the Problem of Nomenclature," *Aliso* 14:125–44.

de Queiroz, K. and M. Donoghue (1988). "Phylogenetic Systematics and the Species Problem," *Cladistics* 4:317–38.

de Queiroz, K. and J. Gauthier (1990). "Phylogeny as a Central Principle in Taxonomy: Phylogenetic Definitions of Taxon Names," *Systematic Zoology* 39:307–22.

de Queiroz, K. and J. Gauthier (1992). "Phylogenetic Taxonomy," *Annual Review of Ecology and Systematics* 23:449–80.

de Queiroz, K. and J. Gauthier (1994). "Toward a Phylogenetic System of Biological Nomenclature," *Trends in Ecology and Evolution* 9:27–31.

de Sousa, R. (1984). "The Natural Shiftiness of Natural Kinds," *The Canadian Journal of Philosophy* 14:561–80.

Dobzhansky, T. (1937). *Genetics and the Origin of Species*. Columbia University Press, New York.

Dobzhansky, T. (1951). *Genetics and the Origin of Species* (3rd ed.). Columbia University Press, New York.

Dobzhansky, T. (1970). *Genetics and the Evolutionary Process*. Columbia University Press, New York.

Donoghue, M. (1985). "A Critique of the Biological Species Concept and Recommendations for a Phylogenetic Alternative," *Bryologist* 88:172–81.

Dupré, J. (1981). "Natural Kinds and Biological Taxa," *Philosophical Review* 90:66–90.

Dupré, J. (1986). "Sex, Gender, and Essence," in P. French, T. Uehling, Jr., and H. Wettstein (eds.), *Midwestern Studies in Philosophy XI*. University of Minnesota Press, Minneapolis, pp. 441–57.

Dupré, J. (1993). *The Disorder of Things: Metaphysical Foundations of the Disunity of Science*. Harvard University Press, Cambridge, Massachusetts.

Ehrlich, P. and P. Raven (1969). "Differentiation of Populations," *Science* 165:1228–32.

Eldredge, N. (1985). *Unfinished Synthesis*. Oxford University Press, New York.

Eldredge, N. and J. Cracraft (1980). *Phylogenetic Patterns and the Evolutionary Process*. Columbia University Press, New York.

Eldredge, N. and S. Gould (1972). "Punctuated Equilibria, an Alternative to Phyletic Gradualism," in T. Schopf (ed.), *Models in Paleobiology*. Freeman Cooper, San Francisco, pp. 82–115.

Enc, B. (1975). "Necessary Properties and Linnaean Essentialism," *Canadian Journal of Philosophy* 5:83–102.

References

Endler, J. (1973). "Gene Flow and the Population Differentiation," *Science* 179:243–50.

Ereshefsky, M. (1988). "Individuality and Macroevolutionary Theory," in A. Fine and J. Leplin (eds.), *PSA 1988*, Volume 1. Philosophy of Science Association, East Lansing, Michigan, pp. 216–22.

Ereshefsky, M. (1989). "Where's the Species?: Comments on the Phylogenetic Species Concepts," *Biology and Philosophy* 4:89–96.

Ereshefsky, M. (1991a). "Species, Higher Taxa, and the Units of Evolution," *Philosophy of Science* 58:84–101.

Ereshefsky, M. (1991b). "The Semantic Approach to Evolutionary Theory," *Biology and Philosophy* 6:59–80.

Ereshefsky, M., ed. (1992a). *The Units of Evolution: Essays on the Nature of Species*. MIT Press, Cambridge, Massachusetts.

Ereshefsky, M. (1992b). "Eliminative Pluralism," *Philosophy of Science* 59:671–90.

Ereshefsky, M. (1994). "Some Problems with the Linnaean Hierarchy," *Philosophy of Science* 61:186–205.

Ereshefsky, M. (1995a). "Pluralism, Normative Naturalism, and Biological Classification," in D. Hull and R. Burian (eds.), *PSA 1994*, Volume 2. Philosophy of Science Association, East Lansing, Michigan, pp. 382–9.

Ereshefsky, M. (1995b). "Critical Notice: John Dupré's *The Disorder of Things*," *Canadian Journal of Philosophy* 25:143–158.

Ereshefsky, M. (1997). "The Evolution of the Linnaean Hierarchy," *Philosophy and Biology* 12:493–519.

Ereshefsky, M. (1998). "Species Pluralism and Anti-Realism," *Philosophy of Science* 65:103–20.

Farris, J. (1976). "Phylogenetic Classification of Fossils with Recent Species," *Systematic Zoology* 25:271–82.

Faye, J. (1994). "What Explains What?" Unpublished manuscript. Center for the Philosophy of Science, University of Pittsburgh.

Feyerabend, P. (1978). *Science in a Free Society*. New Left Books, London.

Frost, D. and D. Hillis (1990). "Species in Concept and Practice: Herpetological Applications," *Herpetologica* 46:87–104.

Futuyma, D. (1986). *Evolutionary Biology* (2nd ed.). Sinauer, Sunderland, Massachusetts.

Gauthier, J. et al. (1988). "A Phylogenetic Analysis of Lepidosauromorpha," in R. Estes and G. Pregill (eds.), *Phylogenetic Relationships of Lizard Families*. Stanford University Press, Palo Alto, California, pp. 15–98.

Ghiselin, M. (1969). *The Triumph of the Darwinian Method*. University of Chicago Press, Chicago.

Ghiselin, M. (1974). "A Radical Solution to the Species Problem," *Systematic Zoology* 23:536–44.

Ghiselin, M. (1987[1992]). "Species Concepts, Individuality, and Objectivity," *Biology and Philosophy* 2:127–43. Reprinted in M. Ereshefsky (ed.), *The Units of Evolution*. MIT Press, Cambridge, Massachusetts, pp. 363–80.

Ghiselin, M. (1989). "Sex and the Individuality of Species: A Reply to Mishler and Brandon," *Biology and Philosophy* 4:77–80.

References

Ghiselin, M. (1995). "Ostensive Definitions of the Names of Species and Clades," *Biology and Philosophy* 10:219–22.

Giere, R. (1988). *Explaining Science*. University of Chicago Press, Chicago.

Goudge, T. (1961). *The Ascent of Life*. University of Toronto Press, Toronto.

Gould, S. J. (1980). *The Panda's Thumb*. W.W. Norton, New York.

Gould, S. J. (1986). "Evolution and the Triumph of Homology, or Why History Matters," *American Scientist* 74:60–9.

Grant, V. (1980). "Gene Flow and the Homogeneity of Species Populations," *Biologisches Zentralblatt* 99:157–69.

Grant, V. (1981). *Plant Speciation* (2nd ed.). Columbia University Press, New York.

Graybeal, A. (1995). "Naming Species," *Systematic Zoology* 44:237–50.

Griffiths, G. (1973). "Some Fundamental Problems in Biological Classification," *Systematic Zoology* 22:338–43.

Griffiths, G. (1974). "On the Foundations of Biological Systematics," *Acta Biotheoretica* 3–4:85–131.

Griffiths, G. (1976). "The Future of Linnaean Nomenclature," *Systematic Zoology* 25:168–73.

Griffiths, P. (1997). *What Emotions Really Are*. Chicago University Press, Chicago.

Guyot, K. (1987). "Specious Individuals," *Philosophia* 37:101–26.

Hacking, I. (1991a). "A Tradition of Natural Kinds," *Philosophical Studies* 61:109–26.

Hacking, I. (1991b). "On Boyd," *Philosophical Studies* 61:149–54.

Hacking, I. (1996). "The Disunities of Science," in P. Galison and D. Stump (eds.), *The Disunity of Science: Boundaries, Contexts, and Power*. Stanford University Press, Stanford, California, pp. 37–74.

Harlin, M. and P. Sunberg (1998). "Taxonomy and Philosophy of Names," *Biology and Philosophy* 13:233–44.

Hempel, C. (1965). *Aspects of Scientific Explanation and Other Essays*. Free Press, New York.

Hennig, W. (1950). *Grundzuge einer theorie der phylogenetischen systematik*. Deutscher zentralverlag, Berlin.

Hennig, W. (1965[1994]). "Phylogenetic Systematics," *Annual Review of Entomology* 10:97–116. Reprinted in E. Sober (ed.), *Conceptual Issues in Evolutionary Biology*, MIT Press, Cambridge, Massachusetts, pp. 257–76.

Hennig, W. (1966). *Phylogenetic Systematics*. University of Chicago Press, Chicago.

Hennig, W. (1969[1981]). *Insect Phylogeny*. Translated by A. C. Pont. John Wiley Press, New York. Originally published as *Die Stammesgeschichte der Insekten*, Waldemar Kramer, Frankfurt.

Heywood, V. (1985). "Linnaeus – the Conflict Between Science and Scholasticism," in J. Weinstock (ed.), *Contemporary Perspectives on Linnaeus*. University Press of America, Lanham, Maryland, pp. 1–16.

Hickman, C., L. Roberts, and A. Larson (1997). *Integrated Principles of Zoology* (10th ed.). McGraw-Hill, New York.

Holsinger, K. (1984). "The Nature of Biological Species," *Philosophy of Science* 51:293–307.

304

References

Horvath, C. (1997). "Some Questions about Identifying Individuals: Failed Intuitions about Organisms and Species," *Philosophy of Science* 64: 654–68.

Hull, D. (1964[1989]). "Consistency and Monophyly," *Systematic Zoology* 13:1–11. Reprinted in D. Hull, *The Metaphysics of Evolution*, SUNY Press, New York, pp. 129–43.

Hull, D. (1965[1992]). "The Effect of Essentialism on Taxonomy: Two Thousand Years of Stasis," *British Journal for the Philosophy of Science* 15:314–26, 16:1–18. Reprinted in M. Ereshefsky (ed.), *The Units of Evolution*. MIT Press, Cambridge, Massachusetts, pp. 199–226.

Hull, D. (1966). "Phylogenetic Numericlature," *Systematic Zoology* 15:14–17.

Hull, D. (1970). "Contemporary Systematic *Philosophies*," *Annual Review of Ecology and Systematics* 1:19–53.

Hull, D. (1974). *The Philosophy of Biological Science*. Prentice-Hall, Englewood Cliffs, New Jersey.

Hull, D. (1976). "Are Species Really Individuals?" *Systematic Zoology* 25:174–91.

Hull, D. (1977). "The Ontological Status of Species as Evolutionary Units," in R. Butts and J. Hintikka (eds.), *Foundational Problems in the Special Sciences*. Reidel Publishing, Dordrecht, pp. 91–102.

Hull, D. (1978[1992]). "A Matter of Individuality," *Philosophy of Science* 45:335–60. Reprinted in M. Ereshefsky (ed.), *The Units of Evolution*, pp. 293–316.

Hull, D. (1979). "The Limits of Cladism," *Systematic Zoology* 28:414–38.

Hull, D. (1980). "Individuality and Selection," *Annual Review of Ecology and Systematics* 11:311–32.

Hull, D. (1985). "Linné as an Aristotelian," in J. Weinstock (ed.), *Contemporary Perspectives on Linnaeus*. University Press of America, Lanham, Maryland, pp. 37–54.

Hull, D. (1987). "Genealogical Actors in Ecological Roles," *Biology and Philosophy* 2:168–83.

Hull, D. (1988). *Science as a Process*. University of Chicago Press, Chicago.

Hull, D. (1989). "A Function for Actual Examples in Philosophy of Science," in M. Ruse (ed.), *What the Philosophy of Biology Is*. Kluwer Academic Publishers, Dordrecht, pp. 309–22.

Hull, D. (1990). "Conceptual Selection," *Philosophical Studies* 60:77–87.

Hull, D. (1992). "The Particular-Circumstance Model of Scientific Explanation," in H. Nitecki and D. Nitecki (eds.), *History and Evolution*. SUNY Press, Albany, New York.

International Association of Microbiological Societies (1992). *International Code of Nomenclature of Bacteria*. American Society of Microbiology, Washington, D.C.

International Botanical Congress (1988). *International Code of Botanical Nomenclature*. Koeltz, Koigstein.

International Commission on Zoological Nomenclature (1985). *International Code of Zoological Nomenclature* (3rd ed.). International Trust for Zoological Nomenclature, London.

References

Jackson, F. and J. Pound (1979). "Comments on Assessing the Differentiating Effect of Gene Flow," *Systematic Zoology* 28:78–85.

Janzen, D. (1977). "What Are Dandelions and Aphids?" *American Naturalist* 111:586–9.

Jeffrey, C. (1989). *Biological Nomenclature* (3rd ed.). Edward Arnold, London.

Jeffrey, R. (1983). *The Logic of Decision*. University of Chicago Press, Chicago.

Khalidi, A. (1993). "Carving Nature at the Joints," *Philosophy of Science* 60:100–13.

Kitcher, P. (1982). *Abusing Science: The Case against Creationism*. MIT Press, Cambridge, Massachusetts.

Kitcher, P. (1984a). "Species," *Philosophy of Science* 51:308–33.

Kitcher, P. (1984b). "Against the Monism of the Moment: A Reply to Elliott Sober," *Philosophy of Science* 51:616–30.

Kitcher, P. (1987). "Ghostly Whispers: Mayr, Ghiselin, and the 'Philosophers' on the Ontological Status of Species," *Biology and Philosophy* 2:184–92.

Kitcher, P. (1989). "Some Puzzles about Species," in M. Ruse (ed.), *What the Philosophy of Biology Is*. Kluwer Academic Publishers, Dordrecht, pp. 183–208.

Kitcher, P. (1993). *The Advancement of Science: Science without Legend, Objectivity without Illusions*. Oxford University Press, Oxford.

Kitts, D. B. and D. J. Kitts (1979). "Biological Species and Natural Kinds," *Philosophy of Science* 46:613–22.

Kornblith, H. (1993). *Inductive Inference and Its Natural Ground*. MIT Press, Cambridge, Massachusetts.

Kripke, S. (1972). "Naming and Necessity," in D. Davidson and G. Harman (eds.), *Semantics of Natural Language*. Reidel, Dordrecht, pp. 253–355.

Lakatos, I. (1970). "Falsification and the Methodology of Research Programmes," in I. Lakatos and A. Musgrave (eds.), *Criticism and the Growth of Knowledge*. Cambridge University Press, Cambridge, pp. 91–196.

Lande, R. (1980). "Genetic Variation and Phenotypic Evolution during Allopatric Speciation," *American Naturalist* 116:463–79.

Lange, M. (1995). "Are There Natural Laws Concerning Particular Species?" *Journal of Philosophy* 77:430–51.

Larson, J. (1971). *Reason and Experience: The Representation of Natural Order in the Work of Carl von Linné*. University of California Press, Berkeley.

Latour, B. and S. Woolgar (1986). *Laboratory Life* (2nd ed.). Princeton University Press, Princeton, New Jersey.

Laudan, L. (1983). "The Demise of the Demarcation Problem," in R. Cohen and L. Laudan (eds.), *Physics, Philosophy and Psychoanalysis: Essays in Honor of Adolf Grunbaum*. D. Reidel Publishing Company, Dordrecht, pp. 111–27.

Laudan, L. (1984). *Science and Values*. University of California Press, Berkeley.

Laudan, L. (1987). "Progress or Rationality?: The Prospects for Normative Naturalism," *American Philosophical Quarterly* 24:19–31.

Laudan, L. (1990). "Normative Naturalism," *Philosophy of Science* 57:44–59.

Lennox, J. (1987). "Kinds, Forms of Kinds, and the More and the Less in Aristotle's Biology," in A. Gotthelf and J. Lennox (eds.), *Philosophical Issues in Aristotle's Biology*. Cambridge University Press, New York, pp. 339–59.

References

Lennox, J. (1993). "How to Find Natural Kinds," unpublished manuscript. Department of History and Philosophy of Science, University of Pittsburgh.

Leplin, J. (1990), "Renormalizing Epistemology," *Philosophy of Science* 57:20–33.

Levin, D. and H. Kerster (1974). "Gene Flow in Seed Plants," *Evolutionary Biology* 7:139–220.

Linnaeus, C. (1731[1938]). *The Critica Botanica*. Translated by A. Hort. The Ray Society, London.

Linnaeus, C. (1737). *Hortus Cliffortianus*. (N.p.), Amsterdam.

Linnaeus, C. (1751). *Philosophia Botanica*. G. Kiesewetter, Stockholm.

Little, F. (1964). "The Need for a Uniform System of Biological Numericlature," *Systematic Zoology* 13:191–4.

Locke, J. (1894[1975]). *An Essay Concerning Human Understanding*. P. Nidditch (ed.). Oxford University Press, New York.

Lovejoy, A. (1959). "Buffon and the Problem of Species," in B. Glass, Q. Temkin, and W. Stauss (eds.), *Forerunners of Darwin*. Johns Hopkins University Press, Baltimore, Maryland, pp. 84–113.

Lovtrup, S. (1977). *The Phylogeny of Vertebrata*. Wiley, London.

Lloyd, G. E. R. (1968). *Aristotle: The Growth and Structure of His Thought*. Cambridge University Press, Cambridge.

Matthen, M. (1998). "Biological Universals and the Nature of Fear," *Journal of Philosophy* 80:105–32.

Mayr, E. (1940). "Speciation Phenomena in Birds," *American Naturalist* 74:249–78.

Mayr, E. (1959). "Typological versus Population Thinking," in *Evolution and Anthropology: A Centennial Appraisal*. The Anthropological Society of Washington, Washington, D.C. Reprinted in E. Mayr, *Evolution and the Divesity of Life*. Harvard University Press, Cambridge, Massachusetts, pp. 26–9.

Mayr, E. (1963). *Animal Species and Evolution*. Harvard University Press, Cambridge, Massachusetts.

Mayr, E. (1965). "Numerical Phenetics and Taxonomic Theory," *Systematic Zoology* 14:73–97.

Mayr, E. (1969). *Principles of Systematic Zoology*. Harvard University Press, Cambridge, Massachusetts.

Mayr, E. (1970). *Populations, Species, and Evolution*. Harvard University Press, Cambridge, Massachusetts.

Mayr, E. (1976). *Evolution and the Diversity of Life*. Harvard University Press, Cambridge, Massachusetts.

Mayr, E. (1981[1994]). "Biological Classification: Toward a Synthesis of Opposing Methodologies," *Science* 214:510–16. Reprinted in E. Sober (ed.), *Conceptual Issues in Evolutionary Biology* (2nd ed.). MIT Press, Cambridge, Massachusetts, pp. 277–94.

Mayr, E. (1982). *The Growth of Biological Thought*. Harvard University Press, Cambridge, Massachusetts.

Mayr, E. (1987). "The Ontological Status of Species: Scientific Progress and Philosophical Terminology," *Biology and Philosophy* 2:145–66.

Mayr, E. (1988). "The Species Category," in E. Mayr, *Toward a New Philosophy of Biology*. Harvard University Press, Cambridge, Massachusetts, pp. 315–34.

307

References

Mayr, E. and W. Provine, eds. (1980). *The Evolutionary Synthesis*. Harvard University Press, Cambridge, Massachusetts.

Mellor, H. (1977). "Natural Kinds," *British Journal for the Philosophy of Science* 28:299–312.

Michener, C. (1963). "Some Future Developments in Taxonomy," *Systematic Zoology* 12:151–72.

Michener, C. (1964). "The Possible Use of Uninominal Nomenclature to Increase the Stability of Names in Biology," *Systematic Zoology* 13:182–90.

Mishler, B. (1990). "Reproductive Biology and Species Distinctions in the Moss Genus *Tortula*, as Represented in Mexico," *Systematic Botany* 15: 86–97.

Mishler, B. and R. Brandon (1987). "Individuality, Pluralism, and the Phylogenetic Species Concept," *Biology and Philosophy* 2:397–414.

Mishler, B. and A. Budd (1990). "Species and Evolution in Clonal Organisms – Introduction," *Systematic Botany* 15:79–85.

Mishler, B. and M. Donoghue (1982). "Species Concepts: A Case for Pluralism," *Systematic Zoology* 31:491–503.

Nelson, G. (1972). "Phylogenetic Relationship and Classification," *Systematic Zoology* 21:227–31.

Nelson, G. and N. Platnick (1981). *Systematics and Biogeography: Cladistics and Vicariance*. Columbia University Press, New York.

Nixon, K. and Q. Wheeler (1990). "An Amplification of the Phylogenetic Species Concept," *Cladistics* 6:211–23.

O'Hara, R. (1988). "Homage to Clio, or, Toward an Historical Philosophy for Evolutionary Biology," *Systematic Zoology* 37:142–55.

Panchen, A. (1992). *Classification, Evolution, and the Nature of Biology*. Cambridge University Press, Cambridge.

Paterson, H. (1985[1992]). "The Recognition Concept of Species," in E. Vrba (ed.), *Species and Speciation*. Transvall Museum, Pretoria, pp. 21–9. Reprinted in M. Ereshefsky (ed.), *The Units of Evolution*. MIT Press, Cambridge, Massachusetts, pp. 139–58.

Patterson, C. (1980[1982]). "Cladistics," *Biologist* 27:234–40. Reprinted in M. Smith (ed.), *Evolution Now*. MacMillan, London, pp. 110–20.

Patterson, C. (1981). "The Goals, Uses, and Assumptions of Cladistic Analysis," presented to the Second Annual Meeting of the Willi Hennig Society, Ann Arbor, Michigan.

Patterson, C. and D. Rosen (1977). "Review of Ichthyodectiform and Other Mesozoic Teleost Fishes and the Theory and Practice of Classifying Fossils," *Bulletin of the American Museum of Natural History* 158:81–172.

Platnick, N. (1979). "Philosophy and the Transformation of Cladistics," *Systematic Zoology* 28:537–46.

Popper, K. (1963). "Science: Conjectures and Refutations," in *Conjectures and Refutations*. Harper and Row, New York, pp. 33–59.

Price, H. (1992). "Metaphysical Pluralism," *Journal of Philosophy* 84:387–409.

Provine, W. (1971). *The Origin of Theoretical Population Genetics*. University of Chicago Press, Chicago.

Provine, W. (1980). "Epilogue," in E. Mayr and W. Provine (eds.), *The Evolu-*

tionary Synthesis. Harvard University Press, Cambridge, Massachusetts, pp. 399–412.

Putnam. H. (1975). *Mind, Language, and Reality. Philosophical Papers*, Volume 2. Cambridge University Press, Cambridge.

Richards, R. (1992). "The Structure of Narrative Explanation in History and Biology," in H. Nitecki and D. Nitecki (eds.), *History and Evolution.* SUNY Press, Albany, New York, pp. 19–54.

Ridley, M. (1986). *Evolution and Classification: The Reformation of Cladism.* Longman, New York.

Ridley, M. (1989). "The Cladistic Solution to the Species Problem," *Biology and Philosophy*, 4:1–16.

Ridley, M. (1993). *Evolution.* Blackwell, Cambridge, Massachusetts.

Rosen, D. (1979). "Vicariant Patterns and Historical Explanation in Biogeography," *Systematic Zoology* 27:159–88.

Rosenberg, A. (1985a). *The Structure of Biological Science.* Cambridge University Press, Cambridge.

Rosenberg, A. (1985b). "Methodology, Theory and the Philosophy of Science," *Pacific Philosophical Quarterly* 66:377–93.

Rosenberg, A. (1990). "Normative Naturalism and the Role of Philosophy," *Philosophy of Science* 57:34–43.

Rosenberg, A. (1994). *Instrumental Biology or the Disunity of Science.* Chicago University Press, Chicago.

Ruse, M. (1969). Definitions of Species in Biology," *British Journal for the Philosophy of Science* 38:225–42.

Ruse, M. (1973). *The Philosophy of Biology.* Hutchinson and Company, London.

Ruse, M. (1976). "The Scientific Methodology of William Whewell," *Centaurus* 20:227–57.

Ruse, M. (1987). "Biological Species: Natural Kinds, Individuals, or What?" *British Journal for the Philosophy of Science* 38:225–42.

Salmon, W. (1984). *Scientific Explanation and the Causal Structure of the World.* Princeton University Press, Princeton, New Jersey.

Salmon, W. (1989). *Four Decades of Scientific Explanation.* University of Minnesota Press, Minneapolis.

Scriven, M. (1961). "The Key Property of Physical Laws – Inaccuracy," in H. Feigl and G. Maxwell (eds.), *Current Issues in the Philosophy of Science.* Holt, Rinehart and Winston, New York, pp. 91–104.

Simpson, G. (1961). *The Principles of Animal Taxonomy.* Columbia University Press, New York.

Simpson, G. (1963). "The Meaning of Taxonomic Statements," in S. Washburn (ed.), *Classification and Human Evolution.* Aldine Press, New York, pp. 1–31.

Simpson, G. (1964). *This View of Life.* Harcourt, Brace and World, New York.

Smart, J. J. (1963). *Philosophy and Scientific Realism.* Routledge and Kegan Paul, London.

Smart, J. J. (1968). *Between Science and Philosophy.* Routledge and Kegan Paul, London.

Sneath, P. and R. Sokal (1973). *Numerical Taxonomy.* W. H. Freeman, San Francisco.

References

Sober, E. (1980[1992]). "Evolution, Population Thinking and Essentialism," *Philosophy of Science* 47:350–83. Reprinted in M. Ereshefsky (ed.), *The Units of Evolution*, MIT Press, Cambridge, Massachusetts, pp. 247–78.

Sober, E. (1984a). "Sets, Species, and Natural Kinds – a Reply to Philip Kitcher's 'Species'," *Philosophy of Science* 51:334–41.

Sober, E. (1984b). *The Nature of Selection: Evolutionary Theory in Philosophical Focus.* MIT Press, Cambridge, Massachusetts.

Sober, E. (1988). *Reconstructing the Past: Parsimony, Evolution, and Inference.* MIT Press, Cambridge, Massachusetts.

Sober, E. (1993). *Philosophy of Biology.* Westview, Boulder, Colorado.

Sober, E., ed. (1994). *Conceptual Issues in Evolutionary Biology* (2nd ed.). MIT Press, Cambridge, Massachusetts.

Sokal, R. (1985[1994]). "The Continuing Search for Order," *American Naturalist* 126:729–49. Reprinted in Sober (ed.), *Conceptual Issues in Evolutionary Biology* (2nd ed.). MIT Press, Cambridge, Massachusetts.

Sokal, R. and T. Crovello (1970[1992]). "The Biological Species Concept: A Critical Evaluation," *American Naturalist* 104:127–53. Reprinted in M. Ereshefsky (ed.), *The Units of Evolution*. MIT Press, Cambridge, Massachusetts, pp. 27–56.

Sokal, R. and P. Sneath (1963). *The Principles of Numerical Taxonomy.* W. H. Freeman, San Francisco.

Splitter, L. (1988). "Species and Identity," *Philosophy of Science* 55:323–48.

Stamos, D. (1996). "Was Darwin Really a Nominalist?" *Journal of the History of Biology* 29:127–44.

Stanford, P. (1995). "For Pluralism and against Realism about Species," *Philosophy of Science* 62:70–91.

Stebbins, G., J. Valencia, and R. Valencia (1946). "Artificial and Natural Hybrids in the Gramineae, tribe Hordeae, I. *Elymus, Sitanion, and Agorpyron*," *American Journal of Botany* 33:338–51.

Templeton, A. (1981). "Mechanisms in Speciation – a Population Genetic Approach," *Annual Review of Ecology and Systematics* 12:23–48.

Templeton, A. (1989[1992]). "The Meaning of Species and Speciation: A Genetic Perspective," in D. Otte and J. Endler (eds.), *Speciation and Its Consequences.* Sinauer, Sunderland, Massachusetts. Reprinted in M. Ereshefsky (ed.), *The Units of Evolution*. MIT Press, Cambridge, Massachusetts, pp. 159–85.

Thagard, P. (1980). "Why Astrology Is a Pseudoscience," in P. Asquith and I. Hacking (eds.), *PSA 1978,* Volume I. Philosophy of Science Association, East Lansing, Michigan, pp. 223–34.

Van Fraassen, B. (1989). *Laws and Symmetry.* Oxford University Press, New York.

Van Valen, L. (1976[1992]). "Ecological Species, Multispecies, and Oaks," *Taxon* 25:233–9. Reprinted in M. Ereshefsky (ed.), *The Units of Evolution*, MIT Press, Cambridge, Massachusetts, pp. 69–78.

Waddington, C. (1957). *The Strategy of the Genes.* Allen and Unwin, London.

Whewell, W. (1840[1984]). "The Philosophy of the Inductive Sciences," in Y. Elkana (ed.), *Selected Writings on the History of Science*. Chicago University Press, Chicago, pp. 121–260.

Wiggins, D. (1980). *Sameness and Substance.* Oxford University Press, Oxford.

References

Wiley, E. (1978[1992]). "The Evolutionary Species Concept Reconsidered," *Systematic Zoology* 27:17–26. Reprinted in M. Ereshefsky (ed.), *The Units of Evolution*. MIT Press, Cambridge, Massachusetts, pp. 79–92.

Wiley, E. (1979). "The Annotated Linnaean Hierarchy, with Comments on Natural Taxa and Competing Systems," *Systematic Zoology* 28:308–37.

Wiley, E. (1981). *Phylogenetics: The Theory and Practice of Phylogenetic Systematics*. Wiley and Sons, New York.

Williams, M. (1970). "Deducing the Consequences of Evolution: A Mathematical Model," *Journal of Theoretical Biology* 29:343–85.

Williams, M. (1985). "Species Are Individuals: Theoretical Foundations for the Claim," *Philosophy of Science* 52:578–90.

Winsor, M. (1985). "The Impact of Darwinism upon the Linnaean Enterprise, with Special Reference to the Work of T. H. Huxley," in J. Weinstock (ed.), *Contemporary Perspectives on Linnaeus*. University Press of America, Lanham, Maryland, pp. 55–84.

Wittgenstein, L. (1958). *Philosophical Investigations* (3rd ed.). Translated by G. E. M. Anscombe. Macmillan, New York.

Index

adaptive zone, 53, 59, 68, 88
anagenesis, 52–3
ancestral characters. *See*
    plesiomorphy
ancestor-descendant relations, 71,
    89, 132, 218, 230
annotated Linnaean hierarchy, 184,
    234, 239–40, 243
"anything goes" objection, 79,
    158–66, 170, 178, 186, 193, 195
apomorphy, 57, 69–71, 75–9, 91–2,
    136, 144, 174–5, 188–9, 260
Aristotle, 16, 18–23, 40–1, 46, 60, 99,
    166
asexual reproduction, 179, 190,
    294n6, 295n1
Atran, S., 296n4
Ax, P., 244–6

basal taxon, 43, 251–5
Beatty, J., 188, 231, 289
Beckner, M., 105
binomials, 204–5, 207, 212–4, 221–4,
    226, 232, 236, 247, 250–5
Biological Species Concept, 35–6,
    81–8, 90, 130, 179, 181, 189, 190–2,
    226, 288
Boyd, R., 27–8,103, 106–8
Brandon, R., 90–1, 132, 134, 147–8,
    153, 191–2
Bryant, H., 298n6

Cain, A., 223, 251–2,
Cartwright, N., 123, 139
causal homeostatic mechanisms,
    27–8, 107–8
causal relations, 15, 28–34, 67, 106,
    110–1, 116–7, 119, 157, 199, 211

causally bound wholes, 29, 113
clade, 71, 259
cladism, 40, 55, 57, 66–70, 73, 75, 79,
    90–2, 192, 195, 199–200, 208–9,
    214–5, 217–8, 220–1, 223–4, 229,
    232, 260–3, 277, 284
    pattern cladism, 51, 75–9, 90–2,
        130, 170, 174–7, 188–9, 192, 219
    process cladism, 51, 66–79, 90,
        132, 170, 173, 175–7, 188, 191
cladogenesis, 52–3
cladogram, 71, 73, 76, 78, 176,
    215–6
cluster analysis, 15, 24–28, 94, 102–3,
    105–7, 213, 259
consilience, 42, 146
Cracraft, J., 92, 173, 177, 215–6, 235,
    243–4
creationism, 47, 158, 199, 205–9, 212,
    214, 232–4, 238–9
Crovello, T., 86–7, 172

Darwin, C., 23, 29, 55, 60, 80, 109,
    208–9, 212, 231, 234, 289, 293n4
Davis, J., 188, 293n7
demarcation criterion, 164–6
de Queiroz, K, 224, 234–5, 241, 253,
    256–62, 266, 273, 279, 287
derived characters. *See* apomorphy
Dobzhansky, T., 52, 81–2, 131, 194,
    293n6
Donoghue, M., 134, 136, 147–8, 153,
    226
Dupré, J., 43, 94, 103, 121, 152–3,
    155, 160–2

Ecological Species Concept, 87–8
Eldredge, N., 60, 96, 115, 117–8, 131,

313

Index

315